WILD
CHINA

WILD CHINA

Text by **JOHN MacKINNON**
Photographs by **NIGEL HICKS**

Produced in association with the World Wide Fund For Nature Hong Kong

The MIT Press
Cambridge, Massachusetts

First MIT Press edition, 1996

MacKinnon, John Ramsay.
     Wild China/ text by John MacKinnon: photographs by Nigel Hicks.
        p. cm.
     Includes bibliographical references (p.   ) and index,
     ISBN 0-262-13329-6 (hardcover : alk. paper)
     1. Natural History — China. 2. Biogeography — China
     3. Natural areas — China.
     4. China — Description and travel. I. Hicks, Nigel. II. Title.
QH181.M33 1996
508. 51 — dc20   96-16953
                                                              CIP

Commissioning Editor: Charlotte Parry-Crooke
Project Manager: Jo Hemmings
Editor: Ann Baggaley
Assistant Editor: Sophie Bessemer
Designer: Behram Kapadia
Page make-up: Sally Kapadia
Cartography: Julian Baker
Index: Janet Dudley

Reproduction by HBM (Singapore) Pte Ltd
Printed and bound in Singapore by Kyodo Printing Co (Singapore) Pte Ltd

CONTENTS

FOCUS ON CENTRAL CHINA *page 144*

FOCUS ON TROPICAL SOUTH CHINA *page 162*

Photographic Acknowledgements

The publishers and photographer extend their thanks to the following people who kindly
loaned their photographs for inclusion in this book. All the photographs in the book,
with the exception of those listed below, were taken by Nigel Hicks.

Aquila (Ray Tipper): page 190 (bottom left); page 196
(top right; centre; bottom left and right); page 197 (bottom).

Ardea (Adrian Warren): page 7; page 76.

Heather Angel: frontispiece; page 33 (top).

China Photo Library Ltd: page 156; page 157.

Bruce Coleman (Erwin and Peggy Bauer): page 33 (bottom); *(Hans Reinhard)*: page 37 (bottom left);
page 111 (bottom); page 133 (top); page 149; *(Rod Williams)*: page 66 (top right); page 111 (top);
(Bruce Coleman): page 129 (left); *(Gerald Cubitt)*: page 148 (top); *(H. Rivarola)*: page 171 (top).

Phillip Cribb: page 27 (bottom).

Tim Loseby: page 152 (centre right).

John MacKinnon: page 21 (bottom); page 29 (top); page 31 (bottom); page 35 (top left;
bottom left and right); page 39 (right); page 40; page 55; page 64; page 65 (bottom right);
page 66 (top left); page 77 (top; bottom left); page 78; page 79 (bottom left and right);
page 81 (bottom left); page 104 (top left); page 105 (top left; centre all three subjects; bottom right);
page 124 (bottom both subjects); page 133 (bottom); page 151; page 154 (bottom);
page 155 (bottom left); page 158 (top); page 160 (both subjects); page 161 (top; centre left; bottom);
page 170 (bottom); page 172 (right, top and bottom); page 173 (left, top and bottom);
page 176 (both subjects); page 177 (bottom left); page 192 (bottom right).

Nature Photographers (E.A. Janes): half title; *(Brinsley Burbidge)*: page 27 (top right); *(W.S. Paton)*:
page 90 (top left); *Hugh Miles)*: page 104 (centre left).

NHPA (Martin Harvey): page 43.

Oxford Scientific Films (Doug Allan): page 35 (top right); *(Zig Leszcynski)*: page 41.

Morten Strange: page 171 (bottom).

Planet Earth (Linda Pitkin): page 200 (top left and right); *(Carl Roessler)*: page 200 (bottom left); *(Mark
Conlin)*: page 200 (bottom right); *(Pete Atkinson)*: page 201 (top); *(Denise Tackett)*: page 201 (bottom).

Wang Sung: page 158 (bottom).

Windrush (David Tipling): page 37 (bottom right); page 82 (all four subjects); page 83 (centre right;
bottom left); page 152 (top; bottom centre); page 197 (top right); *(Goran Ekstrom)*: page 83 (top left and
right; bottom right); page 185 (top right).

WWF (Ron Petocz): page 124 (top); *(Tam Chi Wai)*: page 168 (bottom).

Illustrations appearing in the preliminary pages are as follows:

HALF TITLE: Snow Leopard (*Uncia uncia*).
FRONTISPIECE: Giant Panda (*Ailuropoda melanoleuca*).
PAGE 5: Golden Pheasant (*Chrysolophus pictus*).
PAGE 6: Wild iris in the forests of Tibet.
PAGE 7: The Qinling mountains, north China.
PAGE 10: East coast sunset.
PAGES 12 AND 13: Mount Everest from the Pang La pass, Tibet.

Acknowledgements

The author, photographer and publishers wish to express their thanks to the following for much generous and practical assistance during the preparation of this book:

World Wide Fund For Nature Hong Kong (WWF-HK)

Mary Ketterer • Dr David Melville • Pauline Cheung • Joanna Ruxton •
Llewelyn Young • Catherine Cheung • Florence Lai • Winky So

Particular thanks and appreciation are also extended to:

Dragonair Limited, Hong Kong
Shangri-La Hotels and Resorts, in association with the Shangri-La Hotel, Beijing, the China World Hotel,
Beijing, the Shangri-La Golden Flower Hotel, Xi'an and the Shangri-La Hotel, Hangzhou
The Kadoorie Farm and Botanic Garden, Hong Kong
Jardine Consumer Electronics • Inchcape Pacific Limited, Hong Kong
Bermian Limited, Hong Kong • The Body Shop, Hong Kong

Andrew McAulay and Dr Gary Ades, The Kadoorie Farm and Botanic Garden, Hong Kong
Dr Kenneth Searle and K.K. Lee, Hong Kong Zoological and Botanical Garden
Dr Richard Corlett, Department of Ecology and Biodiversity, Hong Kong
Professor Wang Sung, Chinese Academy of Sciences
Professor Yang Yuanchang, Dr Zhou Xuesong and Liu Jiazhu, Southwest Forestry College, Kunming
Luo Kaiyue and Xia Kairong, Golden Bridge Travel Service, Chengdu
Professor Chen Yiyu and Professor Liu Renjun, Institute of Hydrobiology, Wuhan
Mr Wangdu, Niyty Travel Service, Lhasa
Joe Hsu, Yushan National Park, Taiwan
Kuo Kun-Ming, Kenting National Park, Taiwan
Edward Stokes • Mike Goodwin • Fred, Theresa and Abigail Nilson •
Zhou Haijun• Kasuya Shuji and Takahashi Dai •
Chen Ping • Karen Phillips • Qiu Mingjiang • Vicky Melville •
Mr Laba, Mr Basang and Ms Pozhong •
Monica Lu • Candy Chou

Dr Phillip Cribb, Royal Botanic Gardens, Kew, England
Michael Baron • Craig Robson

The Ministry of Forestry, Beijing

The Bureaux of Forestry of Fujian, Guangxi, Heilongjiang, Hubei, Jiangxi, Shaanxi,
Sichuan and Yunnan Provinces

The staff of the many nature reserves visited, especially at Bawangling, Foping, Hailuogou Glacier Park,
Jianfengling, Jingbo Lake, Jiuzhaigou, Jizushan, Kanas Lake, Poyang Lake, Shennongjia, Taibaishan,
Tangjiahe, Wolong, Wuyishan and Xishuangbanna

The staff of the Xundian branch of the Environmental Protection Agency;
local government officials of Banqiao township (Yunnan);
the staff at the Qiqihar and Jagdaqi branches of the China International Travel Service

Special acknowledgement goes to:

Sir Kenneth Ping-Fan Fung, President, WWF Hong Kong

FOREWORD

China is a land of great beauty and contrasts, ranging from the spectacular snow covered Himalayan peaks to the coastal mangrove swamps, and from dry deserts to tropical rainforests. In this vast country live many animals and plants found nowhere else on earth, including the Giant Panda which has become an internationally recognised symbol of conservation. China, which ranks among the top 10 countries in the world for the richness of its flora and fauna, is a key country for WWF's global conservation efforts.

This vast land, covering some 7% of the land area of the planet, also supports nearly a quarter of the world's human population. As a result, the pressures on its natural resources are immense, especially in the eastern third of the country where some two-thirds of the population live.

China is currently experiencing rapid economic development which is further straining the environment. Pollution of both air and water are of growing concern, as is the need for increased energy production, and the depletion of freshwater resources.

While conscious of the need to balance the many human and other requirements, major efforts are being made to safeguard China's remaining natural resources. Large numbers of nature reserves have been established, and conservation is now accorded increasing importance at both national and provincial levels. WWF is working closely with various government agencies to promote wildlife conservation, as well as addressing many broader environmental issues.

The size of the country, and the remoteness of many of the areas which are blessed with great natural beauty and rich in wildlife, will deter all but the most determined traveller. *Wild China*, however, allows us all to tour the country and learn of both its natural wonders and the many problems it faces, resulting in an increased appreciation and understanding of China's natural environment.

SIR KENNETH PING-FAN FUNG
PRESIDENT
WWF HONG KONG

INTRODUCTION

China is a vast country comprising 10 million square kilometres (4 million square miles), or 5 per cent of all the land on this planet. Such is the complexity of topography and climate in this immense area that, despite its northerly latitudes, it is biologically one of the three richest countries in the world. However, this wealth has not been on full display to western eyes, but has been hidden by centuries of remoteness, mystery and cultural isolation. China is the oldest continuous civilization on earth, yet throughout much of its history was almost unknown to Europe, affecting early European developments only indirectly through the long trade routes travelled from the East.

In 1295 the famous Venetian explorer Marco Polo returned to his native city after a trip of 17 years travelling through the Orient. He wrote the *Book of Marvels* about his experiences, claiming to have visited China. This amazing journal describes the imperial court of the great Kublai Khan at the height of the Yuan dynasty, some 50 years after his grandfather, the Mongol warrior Genghis Khan, had driven the Chinese Song dynasty leaders from their refuge in southern China and established a lineage in the new northern capital near Peking (Beijing). European readers must have raised their eyebrows at the tales of immense wealth, the gilded plates, fine silks, court elephants, paper money, noodles, coal-burning and other wonders. Even more astonishing were the reports of fantastic landscapes, endless deserts, great forests, snow-capped mountain ranges of unbelievable height and strange wildlife – rhinoceroses, peacocks, gibbons, splendid deer, huge snakes and gorgeous pheasants.

Many regarded Marco Polo's account as too fanciful and exaggerated to be true – possibly with good reason, as today some scholars are of the opinion that the intrepid merchant never reached China at all but based his claims on the observations, or even inventions, of other travellers. Certainly no one has ever seen the giant birds that carry elephants into the sky to dash them onto the rocks beneath and feast on their bones – maybe this was an embellishment introduced by Polo's chronicler, the romantic writer Rustichello – but the fact remains that almost everything else has since been verified.

Far from opening up the East, Marco Polo's findings, authentic or not, merely placed Cathay in the realms of mystery and China retained its isolation for a few more centuries. Biological exploration was left to Chinese artists and poets and it was not until the nineteenth century that European missionaries were able to bring back detailed maps, descriptions and specimens of China's natural wonders.

China proved a treasure trove for visiting western naturalists. The French missionary Père Armand David, who was based at Muping in Sichuan province but travelled widely through China, was in the enviable position of naming several hundred mammals, birds and plants new to science. Perhaps his most notable 'discovery' was the Giant Panda, or 'iron-eating beast' as one of its old Chinese names describes this now famous creature.

At the turn of the century the British ornithologist La Touche was also able to find many new species in south-east China. Other expeditions sponsored by western museums and institutions started documenting the north and west of the country. The British botanist E.H. White, collecting plants in south-west China at the turn of the century, like Augustine Henry before him, reaped many exciting plant finds. He brought back some three tons of seeds of azaleas, rhododendrons, evergreen shrubs and other plants that now form the basis of almost every 'English' garden. Of the 3,000 plant species brought back by White over a thousand were new to cultivation in the West.

However, this 'open door' period in the late Qing dynasty was also to end long before the country was fully explored. The Japanese War (1931–45), followed by the Second World War and the Cultural Revolution of the late 1960s put an end to further western involvement.

Russian scientists, in the tradition of the great nineteenth-century explorer Nikolai Przewalski, assisted China in the post-revolutionary period to establish a large Academy of Sciences and it was left to the national scientists to complete the detailed inventory and description of thousands of extra species and varieties. Cut off by the barriers of language and politics, few in the West knew much about these new discoveries. It was not until the 1980s that once again the veil was lifted and western biologists were able to explore China's wild places, and even then with considerable restrictions. When George Schaller started studies of the Giant Panda in 1981, under agreement between the World Wide Fund For Nature (WWF) and the Chinese government, he was reprimanded for taking incidental notes on the rare Golden Monkeys because these were not specified in the study agreement.

Today, however, the door is really open and foreigners can see the country in a way never possible before, from the heights of the Himalayas and vast expanses of the Tibetan plateau to the distant mountains of Tianshan and Altai in the extreme north-west, the great Gobi desert and Loess plateau, complex mountain ranges of central and south-west China, the wild forests of Manchuria in the north-east down to the tropical islands of the south coast. But China's wildernesses will not last for ever. With a fifth of the world's population enjoying the fastest current economic growth, progress and development are carving into what is left of the natural environment at an unprecedented rate. The opening up of the economy has made items that once belonged to nobody suddenly of new value. The pursuit of money is no longer forbidden in China and the insatiable appetite of the huge population for building timber, firewood, fish, water, soil, medicinal plants and anything even remotely

OPPOSITE PAGE The stunning Pearl Shoals falls are the centrepiece of Jiuzhaigou Nature Reserve, a region in northern Sichuan famous throughout China for its steep snow-capped mountains, dense forests, waterfalls and lakes.

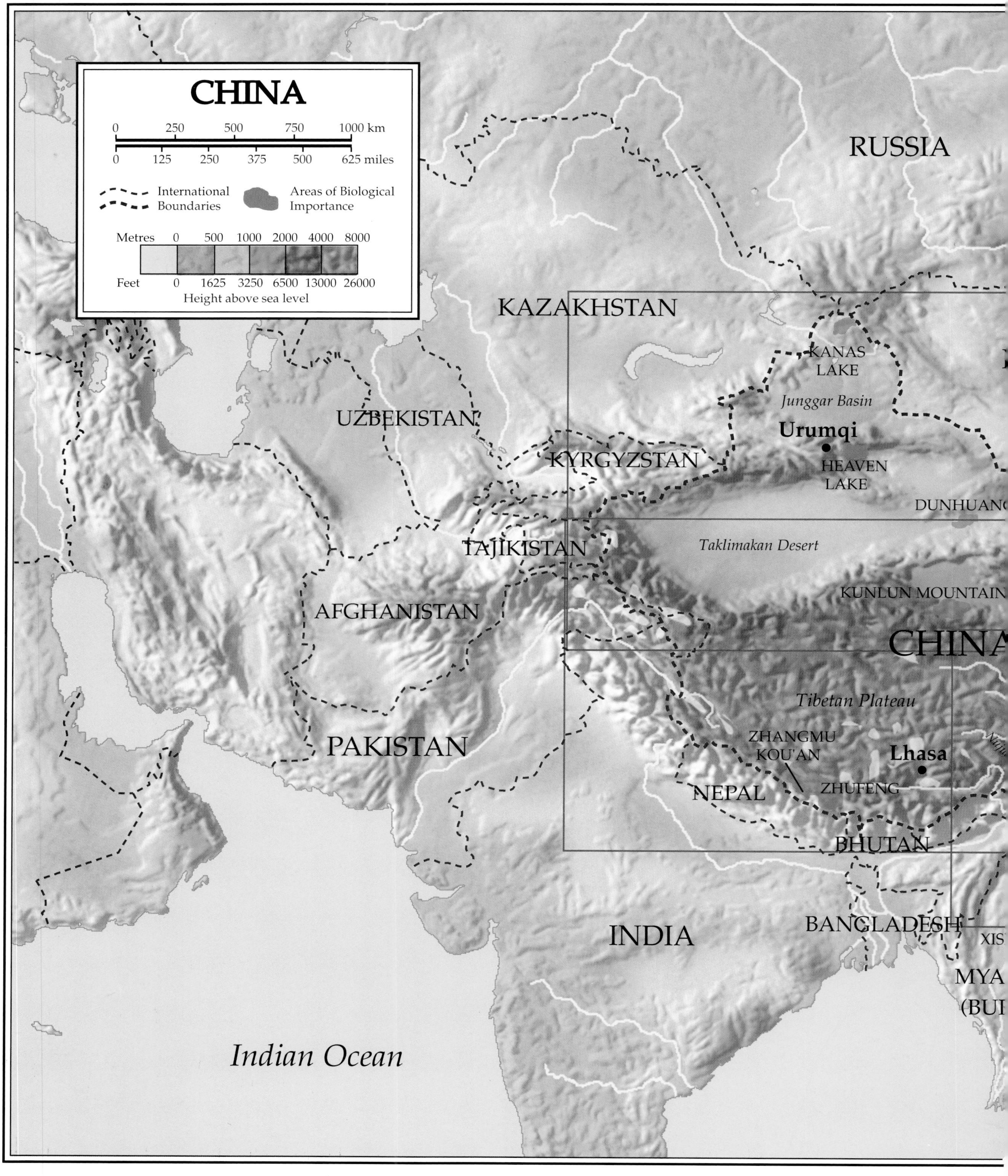

CHINA

0   250   500   750   1000 km

0   125   250   375   500   625 miles

International Boundaries

Areas of Biological Importance

Metres	0	500	1000	2000	4000	8000
Feet	0	1625	3250	6500	13000	26000

Height above sea level

RUSSIA

KAZAKHSTAN

UZBEKISTAN

KYRGYZSTAN

KANAS LAKE

Junggar Basin

Urumqi

HEAVEN LAKE

DUNHUANG

Taklimakan Desert

KUNLUN MOUNTAIN

TAJIKISTAN

AFGHANISTAN

CHINA

Tibetan Plateau

ZHANGMU KOU'AN

Lhasa

PAKISTAN

NEPAL

ZHUFENG

BHUTAN

INDIA

BANGLADESH

XIS

MYA
(BUI

Indian Ocean

N

Heilongjiang (Amur)

ZHALONG
CRANE RESERVE

Harbin

JINGBO
LAKE

CHANGBAISHAN

NORTH
KOREA

JAPAN

WULINGSHAN

Beijing

Hohhot

FRAGRANT
HILLS

BEIDAIHE

Tianjin

SOUTH
KOREA

ONGOLIA

INGHAI
LAKE

TAIBAISHAN/
QINLING MOUNTAINS

Huanghe (Yellow)

Zhengzhou

Shanghai

East
China
Sea

Lanzhou

Xi'an

JIUZHAIGOU

Hangzhou

Wuhan

HUANGSHAN

SHENNONGJIA

WOLONG

Chengdu

EMEISHAN

Yangtze (Changjiang)

YANGTZE
RIVER THREE
GORGES

POYANG
LAKE

Nanchang

Fuzhou

WUYISHAN

Chongqing

Changsha

TAROKO

YUSHAN

TAIWAN

Pacific Ocean

GUANGXI KARST
LIMESTONE LANDSCAPE

YULONG
XUESHAN

Guilin

KENTING

HABA
XUESHAN

Kunming

Guangzhou

MAI PO MARSHES

ANGBANNA

Nanning

HONG
KONG

MAR

LAOS

HAINAN

WUZHISHAN

BAWANGLING

JIANFENGLING

South
China
Sea

THAILAND

(A)

VIETNAM

edible or saleable is threatening almost every living species. Huge dams and other development projects are reshaping landscapes of great antiquity and the clear blue skies that once characterized the country are becoming replaced by a pall of dust and smog.

PHYSICAL GEOGRAPHY

Physically, China can be divided into three major realms. The highest consists of the Tibetan (Qinghai-Xizang) plateau which slopes slightly from a base elevation of about 5,000 metres (16,000 feet) in the west to about 4,000 metres (13,000 feet) in the east. Along the south wall of this plateau the great Himalayas rise much higher, with many peaks over 7,000 metres (23,000 feet) and the world's highest peak, Mount Everest (Qomolangma), at 8,848 metres (29,029 feet). On the northern flank of the plateau, from west to east,are the Kunlun, Altun and Qilian mountains.

Due to the high altitude, temperatures over the entire plateau are low, permafrost is widespread, solar radiation is intense and winds are very strong. Rates of desiccation and erosion are high and the water that comes from snow and glacier melt-off gathers in salty lakes scattered over the plateau. Soils are weak and generally infertile.

Draining from the plateau are five of the world's major rivers. In the south is the Yarlungzangbo or upper stretches of the great Brahmaputra. Next is the Nujiang or Salween followed by the Lancangjiang or Mekong and then the famous Yangtze (which the Chinese call Changjiang). All these rivers start by flowing east but turn steeply south on reaching the eastern end of the Tibetan plateau. The last three run side by side in parallel courses for many kilometres before their diverging paths take the

Salween south-west through Myanmar (Burma), the Mekong south through Laos to Cambodia and Vietnam, and the Yangtze east to wind through the centre of China to the east coast near Shanghai. The fifth river, the Yellow or Huanghe, rises in the north-east of the plateau and drains east across northern China.

At the northern end of the plateau is the curious Qaidam basin which drops down to altitudes of only 2,600 metres (8,530 feet). This is a graben, or area of tectonic collapse. At the bottom of the depression are collected concentrations of minerals. There are salt lakes and this is one of the world's richest salt-mining areas, where even the roads are paved with solid blocks of salt.

Lake Qinghai (Koko Nor), at an elevation of 3,200 metres (10,500 feet) in the north-eastern part of the plateau, is China's largest lake with a surface of 4,400 square kilometres (1,700 square miles). However, like other lakes in the region, its water level is dropping.

The second major realm of physical China is the arid north-western region. In fact this is the eastern end of the great Eurasian desert and grassland zone. It accounts for about 30 per cent of the land area of China but supports only about 5 per cent of the human population.This region is a series of low-lying basins mostly between 500 and 1,000 metres (1,640–3,280 feet) above sea level. As a result of the great distance from any seas and the surrounding ranges of mountains the influence of

BELOW Bogda Feng, at 5,445 metres (17,864 feet), is the highest peak in the eastern limits of the Tianshan mountains that cut across the northern part of Xinjiang in China's far north-west.
OPPOSITE PAGE The Chingshui cliffs in Taiwan's Taroko National Park provide a spectacular setting for the island's Pacific coast. Here, the mountains plunge from 2,500 metres (8,200 feet) straight into the ocean, reaching a trench over 4,000 metres (13,120 feet) deep just a few kilometres offshore.

ABOVE An autumnal view of mixed birch and oak forest in Wulingshan Nature Reserve, located in the mountains north-east of Beijing, straddling the border with Hebei. The reserve is famous as a site for a few wild Ginkgo trees as well as home to the world's most northerly Rhesus macaques.

BELOW Coniferous forests of the Da Hinggan mountains, a range that forms China's northernmost boundary with Mongolia and Russia. China's north-east, once known as Manchuria, is still one of the country's most extensively forested regions, from the broadleaved forests of the south to the larches of the northerly permafrost zone.

ABOVE The Whispering Sands, mighty 250-metre (820-foot) high dunes that crowd right up to the edge of the Dunhuang oasis in Gansu province. The deserts of China's north-west are among the world's largest and most desolate, greatly feared by travellers since the heyday of the Silk Road.

BELOW Extensive grasslands around Lake Kanka, Xingkai to the Chinese, a massive wetland straddling the Sino-Russian border in the far north-east. The plains of north-eastern China are a natural mix of grasslands, marshes and lakes, a habitat vital to tens of thousands of migratory birds.

summer monsoons is very weak and conditions range from arid in the west to semi-arid in the east.

Two great mountain ranges rise out of the north-west – the Tianshan and Altai. Both are glaciated on their highest peaks and the streams that drain off their slopes provide an important but narrow fringe of fertility around the edges of the sandy deserts they separate.

Largest and fiercest of these deserts is the huge Taklimakan immediately to the north of the Tibetan plateau, with an area of 327,000 square kilometres (126,000 square miles), lying in the bottom of the Tarim basin. Its Uigur name means 'once you enter, you can never get out'. Most of the desert comprises loose yellow sand dunes. These dunes, or barkhans, are crescent-shaped and are driven slowly southwards through wind action. The sands are stabilized around the desert's edges and along some river beds by belts of tamarisk bushes. This is the only warm temperate desert in China. Beneath the sands lie buried tracks and oases of the 'Silk Road' where ancient traders led their camel caravans over 2,000 years ago.

To the north of the Tianshan mountains lies another large basin, the Junggar. At 47,000 square kilometres (18,000 square miles), this is China's second largest desert, but quite different from the Taklimakan. Because it is cooler, rates of evaporation are lower and rainfall is slightly greater which allows for more vegetation. Moving sands form the central parts but around the edge of the sands are stabilized stony deserts with enough plant life to support winter grazing.

Smaller deserts include the Alxa in western Inner Mongolia between the Helan and Qilian mountains, and the Ordos south of the great bend of the Yellow river. The Alxa includes the Badain Jaran, Tengger and Ulan Buh desert areas. Mobile dunes cover much of this region but there are some saline lakes and springs, and even some fish living in the Tengger. These deserts are spreading towards the Yellow river but major efforts to afforest and stabilize dunes with grasses are having some effect in checking the desertification process.

To the north, the great stony Gobi desert stretches into the heart of Mongolia, mostly outside Chinese territory. To the east the deserts give way to semi-desert and temperate steppe grasslands as the climate becomes semi-humid.

The third major physical realm is eastern monsoon China, comprising about 45 per cent of the country and containing 95 per cent of the human population. Almost all potentially arable land has already been developed, often overused, degraded or even abandoned, and almost all natural vegetation has been modified by man.

This entire region enjoys seasonally humid weather but there are major seasonal variations in temperature and wind direction, and a gradual cline of climate from the colder temperate north through the subtropical south of the country to a narrow tropical fringe in the furthest south. Landform consists of wide alluvial valleys of the main rivers – particularly the lower reaches of the great Yellow river and the Yangtze – coastal plains, and a few rather ancient and stable ranges of mountains, rarely exceeding 2,000 metres (7,000 feet).

A major fault line cuts across China from south-west Tibet to the north-east of the country. This is the junction of two continental plates and, from time to time, lateral movements of one plate against the other cause earthquakes. Every year a few quakes are recorded and when these are centred on populated areas terrible damage can result. In 1920 180,000 people died when a quake hit Gansu and in 1976 many lives were lost and tens of thousands injured in a series of quakes in the Tangshan industrial region of Hebei province.

The Natural Regions of China

China can be divided into seven major biogeographical units: **North-east**; **North**; **North-west**; the **Tibetan Plateau** and the **Himalayas**; **South-west**; **Central**; and **Tropical South China**. Each has its own distinctive characteristics of landform, vegetation and wildlife, which are described in detail in the Focus sections of this book. Within each section, following the most logical scheme of travel, selected locations of special biological importance and interest are highlighted.

CLIMATE

Climate varies vastly over the face of China, the greatest influence being exerted by the pattern of monsoon winds over the Asiatic landmass. In the winter months, Asia is much cooler than the Pacific Ocean and a high-pressure area is formed. This results in cold dry winds that sweep from the interior landmass south-east towards the ocean. Most of eastern China is swept by this dust-laden monsoon and only a little rain comes from the eastwards-moving depressions to fall in the central regions. In summer, however, Central Asia heats up and in a reverse pattern the monsoon blows from the Pacific and Indian Oceans, bringing with it most of China's annual rainfall.

The north-west and west of China are characterized by low rainfall and very extreme summer high and winter low temperatures. Eastern China is more humid, equable and more suitable for agriculture. In Harbin, in the north-east, the average temperature in January is -18°C (-0.4°F) rising in July to 22°C (72°F), a difference of 40°C (72°F). By contrast, in Guangzhou, in the south of China, the mean January temperature is 13°C (55°F) and in July is 22°C (72°F), a difference of only 9°C (16°F). In the north-east, the plain is snowbound all winter, whilst in the south frost occurs only on the higher hills. Between these extremes the Yangtze valley averages about 3°C (37°F) in mid-winter.

BIOLOGICAL RICHNESS OF CHINA

China's immensity and diversity make it a veritable storehouse for a large proportion of the world's faunal and floral species. For instance, the country supports 500 species of mammals and over 1,200 species of birds, both groups representing 13 per cent of their global totals, while fish make up an equally impressive 12 per cent of all known species. The 25,000 species of higher plants identified in China comprise 11.4 per cent of the global total and numbers for fungi (17 per cent) and algae (19 per cent) are even higher. One group that is especially well represented in China is that of the Gymnosperms, which include the conifers. There are 200 species of these trees in China out of a total of only 520 species known worldwide.

The insect group alone falls short of these extraordinary percentages. China has documented over 40,000 insect species – which sounds a great many but is, in fact, only 5 per cent of the global total. However, their study is still far from complete and their numbers may rise several times over when the tiny tropical beetles become fully accounted for.

All in all, these figures show that China is enormously important for mankind in preserving biological diversity or the genetic building blocks on which all future advances in domestication of wild species, agricultural crop improvement, continued protection of crops and stocks against disease, and a lot of medicinal advances will depend.

Not only does China contain many species which also occur in neighbouring countries but a high proportion of these species are endemic, or confined to China alone. Levels of endemism vary from group to group but, for instance, some 15 per cent of China's mammals, 8 per cent of its birds, 7 per cent of its reptiles, 11 per cent of amphibians, 16 per cent of fish and 7 per cent of higher plants are found nowhere else. Some of these species have already proved to be very important. Silk moths have been farmed for over a thousand years and silk was one of China's major exports for centuries. Silver Carp (*Hypophthalmichthys molitrix*), Snakehead Fish (*Aristichthys nobilis*) and Grass Carp (*Ctenopharyngdon idellus*) are now well-known aquaculture fish all over the world, even though paradoxically they are almost extinct in their original wild habitats. Similarly yaks, camels and wild horses were domesticated here for many centuries. Such trees as Chinese Fir (*Cunninghamia lanceolata*) and Chinese Pine (*Pinus massoniana*) are globally important in silviculture. Perhaps the most significant examples of biodiversity in China are the wild foods. Many varieties of wild rice (*Oriza sativa, O. meyeriana, O. officianalis*) as well as soyabean, barley, tea, apples, beans, kiwi fruit and a wide range of other food species originated in China and many more remain undiscovered or under-utilized. Several thousand more species of plants are used by local people as foods or medicines as well as for fibre, dyes and ornaments.

Little of north China's original forest vegetation now remains undisturbed, but in the shelter of Wulingshan Nature Reserve, to the north-east of Beijing, oaks (*Quercus*) such as this still survive.

China's Native Flora

The vegetation of China, predominantly temperate and subtropical, is extremely ancient and very rich.

Forests naturally dominate in the mountain ranges and eastern hills. In the extreme north-east and the ranges of Altaishan and Tianshan in the north-west, cold conifer forests are the norm, as they are also in the subalpine zone of the higher mountain ranges of the central and south-west regions. Common genera include firs (*Abies*) on the higher ridges, spruces (*Picea*) and hemlocks (*Tsuga*) in the main coniferous blocks and pines (*Pinus*) on drier soil, mixed with oaks (*Lithocarpus*) and chestnuts (*Castanopsis*) in the interface zone between conifer and broadleaf forests. Larches (*Larix*) are common in the permafrost zone of the north-east, on some mountain summits and as fringing trees along riverbeds and swamps. Junipers (*Juniperus*) and yew trees (*Taxus*) occur in the highest montane zones, often in formation with rhododendrons which abound as an understorey. The moister conifer forests also tend to have dense stands of subalpine bamboos as an understorey.

In the subtropical zone are found natural stands of the Chinese Fir. A much rarer and more restricted conifer is the endemic Dawn Redwood (*Metasequoia glyptostroboides*), which occurs naturally only in a small area of southern Hubei but which has also now been used extensively in silviculture and as a roadside tree over much of the subtropical zone. *Metasequoia* is not as tall as its famous relatives the North American redwoods but is a fine stately tree with a delicate pinnate leaf. Like larch it is a deciduous conifer, its leaves turning a gorgeous golden colour before they are shed in winter.

In many parts of China, villagers tap conifer trees for their resins, which are used in making paints, varnish and turpentine. Pines have the most valuable resins and also produce edible seeds. In the north-east, the seeds of Red Pine (*Pinus koraiensis*) are commonly collected for food whilst in the south-west the smaller seeds of White Pine (*P. armandi*) are also eaten both raw or cooked.

Immediately below the conifer zone, in terms of both altitude and latitude, are the mixed deciduous temperate broadleaf forests. These are rich with oaks (*Quercus*), chestnuts (*Castanea*), beeches (*Fagus*), limes (*Tilia*), planes (*Platanus*), aspens (*Populus*) and maples (*Acer*). Similar to the broadleaf forests of Europe or the east coast of the United States, but much better endowed with genera and species, they contain a wealth of edible plants such as plums and cherries, wild pears and apples, gooseberries, currants and brambles. They are also the source of large numbers of herbs used in Chinese traditional medicine. Here, the fruits of Chinese hawthorns (*Crataegus*) grow plump and large, and are collected for medicines. In autumn, the deciduous forests take on brilliant tints, scarlet maples setting off the golden aspens and browner beeches and oaks. The forest floor is covered with ferns and saplings, and colourful fungi feasting off the rich debris of rotting wood and leaves.

Below the temperate zone are the subtropical forests. These are also broadleaf and dominated by oaks (*Lithocarpus* and *Cyclobalanopsis*) and *Castanopsis* chestnuts, but many different species make up the canopy, and the phenology of the trees and form of the leaves (which are thicker, hard and often shiny) are quite different from those of temperate forests. In moister regions these forests are evergreen but in some seasonally dry eastern areas they may be largely deciduous. This deciduous nature is a response to a hot and dry climate not to the cold frosts which trigger the leaf loss in temperate forests.

Some of the subtropical forests, although not lacking in species diversity, appear very uniform in structure. This is where several species of *Cyclobalanopsis* oaks and *Castanopsis* chestnuts have evolved to convergent form, and one can be fooled into thinking the forest is a monoculture until one looks in closer detail at the minutiae of leaf and fruit forms. In fact, these forests contain many rare and interesting species including several of the ancient genera that once predominated in the prehistoric forests of the Tethys depression. The beautiful Dove-tree (*Davidia involucrata*) named after the shape of its white flowers, and the rare Spur-leaf (*Tetracentron sinense*) are examples. Other important trees and bushes include the Sichuan Pepper Tree (*Zanthoxylum simulans*), the Wild Pomelo (*Citrus grandis*) and many lovely magnolias, camellias and azaleas. One species of camellia (*Camellia sinensis*) was domesticated centuries ago as the ancestor of tea.

In addition, the undergrowth and secondary scrub of this zone are home to many horticultural plants and shrubs familiar to gardeners in the West: purple-flowering buddleias that attract hosts of butterflies, yellow-flowered berberis bushes with spiny holly-form leaves, red-flowered peonies and yellow clumps of chrysanthemums.

Two-thirds of the land area of China is montane and such a great diversity of mountains provides varied landforms and ecological conditions for a vast number of plants to grow and survive. The harsh conditions – cold, wind, high insolation and poor soils – place constraints on the lifeforms that can survive there but, within these limits, the plant kingdom is quite amazing and the variety of forms and beauty of many of the flowers is fantastic. This alpine flora also forms an important source of materials of proven medicinal value.

The alpine zone is well defined as the treeline is generally quite abrupt. Forests of fir and hemlock suddenly give way to scrub of juniper and rhododendron and above that spreads a patchwork of heath (generally composed of dwarf species of rhododendron and a few other alpine bushes), meadows and pastures which have a high proportion of grass but also many other herbs and shrubs. Several plant genera are particularly well represented in the alpine flora.

In the subalpine forest understorey many rhododendron species grow as large bushes with showy flowers but out of the forest, fully exposed to wind and frost, they are heather-like with rather small simple flowers. Some have tubular red flowers, which are pollinated by sunbirds, but many have pink, purple or even yellow blooms. In treeless areas these bushes may be the main source of firewood and in such places the rhododendron scrub may be seriously overexploited. Other common shrubs are *Potentilla*, *Cotoneaster* and *Hypericum*.

Poppies are another familiar group in the alpine flora. The beautiful blue *Mecanopsis horridula* is one of the more widespread species but varies greatly in form depending on exposure. In some areas it is tall and elegant, in others crouching and hairy. Other species may be yellow or pink. Most famous in China is the Opium Poppy (*Papaver somniferum*) which was cultivated in Sichuan as early as the Tang dynasty (AD 618-907) with seed imported from Persia.

Beautiful blue gentians open their trumpet flowers in bright sunshine. Some have small branched flowers but others are more robust and splendid. Louseworts (*Pedicularis*) are extremely widespread and form colourful pink or purple clumps, whilst the delicate flowers of *Corydalis* are an electric pale blue. Geraniums, primulas, anemones, potentillas, vetches and tiny strawberries are all very common with many representative species. Asters open boldly with rays of fine purple petals fringing their daisylike yellow centres. Distinctive purple irises flower in the taller herb areas.

In the alpine zone are found several species of wild rhubarb, with broad, crinkly triangular leaves and tall stems with many tiny red flowers. One species, *Rheum nobile*, has great white bractlike leaves that hang like handkerchiefs and even at a long distance can be seen standing tall above the other herbs.

One unusual group of alpine plants are the *Sassaurea*. These fleshy, thistle-like flowers have hairy or down-covered bracts around the flowerheads which help protect them against the cold. The leaves, too, are thermally insulated with dense hair.

In the sheltered stream-beds and thickets grow alpine willows and pink-flowering wild roses, with thistles, nettles and bushes of blue-flowering *Sophora* and *Berberis*. Alpine grasslands are dominated by *Kobresia* and *Stipa* species whilst in marsh areas *Blysmus* and *Carex* may be common. Both are important pasture for wild animals and domestic herds.

The tropical zone of China supports the richest vegetation in the country, particularly in the low-lying evergreen forests of south-west Yunnan and Hainan. Such forests can be very tall, especially in the small patches of dipterocarp forest dominated by *Parashorea* and *Vatica* trees that tower on straight unbranching boles for over 30 metres (100 feet) before branching into compact rounded crowns that form a canopy ceiling of up to 50 metres (160 feet).

Rainforests are the most complex vegetation type in terms of structure, with various strata identified as emergent, canopy, middle, lower and understorey. There are several hundred species of tree in each square kilometre and a much larger number of herbs, bushes, epiphytes, twisting lianas, ferns, mosses and other lower plants.

The great steppes of northern China are characterized by the dominant grass genus *Stipa*, but there are many flowering herbs such as this elegant iris (*Iris scariosa*) in Inner Mongolia.

The temperate broadleaf forests of north and east China are dominated by oaks (*Quercus*) but white-barked birches (*Betula*) are the main infillers and colonizers.

The Dawn Redwood (*Metasequoia glyptostroboides*) is a deciduous conifer endemic to a small area of central China, but is now grown the world over as a popular silvicultural tree.

Many of the trees have huge glossy leaves; for instance, those of *Magnolia henryi* are up to a metre (3 feet) long. Some have fin-like flaring buttresses to give them greater support on the generally shallow soils and the larger trees are usually covered in epiphytic orchids and ferns, a bewildering variety of which still defy description and are largely unnamed. On the forest floor grow great clumps of wild gingers with gaudy white or red flowers and leaves that smell aromatic and fragrant when crushed. The giant 'elephant-ear' leaves of wild yams, often riddled with the perfect circular holes made by insects, are strikingly noticeable. Fungi abound in the damp litter, including the strange-looking and foul-smelling basket fungi.

Dominant tree families of the lowland rainforest are Lauraceae, Moraceae, Rutaceae and Myristicacae. There are many edible fruits including figs, wild persimmons, wild mangosteens and tiny sweet mangoes. These provide much of the food of gibbons, monkeys, hornbills and pigeons, but can also be eaten by human visitors to these dark damp forests.

Palms are common in the tropical zone. Around the coast are groves of tall coconuts but these are all introduced. The tallest native palms are the elegant *Livistona* trees with round fan-leaves that are used for roofing and, when young, as a local version of the carrier bag. Another important genus of palms is *Caryota*. This is one of the so-called 'fish-tail' palms with a complex leaf structure of triangular sections on springy raylike stems. *Caryota* palms are one of the favourite foods of elephants and the flower-stems can be cut to trap the rich sugary sap which minority farmers ferment to make alcoholic toddy. Largest and most splendid is the rather curly-leaved species *C. urens*, which is particularly found in association with limestone regions.

Other smaller palms include clumps of *Licuala*, whose split fan-leaves terminate so abruptly they look as though they have been trimmed with scissors. Slender *Pinanga* palms grow in moist rainforests, whilst the spiny *Phoenix* palms grow in dry sandy areas along the coastline. Climbing rattans (*Calamus*) are confined to dense rainforests and use their hooked leaf-spines to cling to other bushes and trees as they climb up to the canopy to flower. Their strange scaly fruits are eaten by monkeys and even knowing farmers.

Great expanses of grassland and desert extend across the arid heartlands of Eurasia from Europe to the Amur river in the east, constituting one of the world's most extensive vegetation types. Within China such habitats cover half of the country's landmass with deserts being centrally and westerly surrounded by stony gobi and lusher grassy steppes, finally grading into woodland meadows. Throughout history the grasslands have been enormously important to China as the home of the horseman warrior, the easy communication routes to Central Asia and for the rearing of domestic stocks. Today, much of this natural grass area has already been converted to grain cultivation.

The grasslands of China vary greatly in species composition determined by factors of climate, water tables and soil types. Major grass groups include *Stipa*, *Arundinella*, *Festuca* and *Aneurolepidium*, with some *Acantherum*, *Agropyron*, *Eragrostis* and other genera involved to a lesser extent. *Acantherum* is a coarse steppe grass which forms dense clumps. The many species of *Stipa* abound on the plains and many herb genera are associated with these grasses, with Leguminosae and Compositae families much in evidence. However, the most important herb genus is undoubtedly *Artemisia* – an aromatic group of plants, many species of which are regarded as medicinal.

Where scattered trees become associated at transition zones of grassland it is usually the drought-resistant elm *Ulmus pumila* which is first to appear, especially in slightly saline areas. Other grassland trees include poplars (*Populus*), Mongolian Willow (*Salix mongolica*) and Mongolian Oak (*Quercus mongolica*).

In winter the steppes are cold and wind-blown but in the early summer they are lush and green, humming with the wings of insects and decorated with the flowers of irises, asters, violets, veronicas, wild onions, polygonums and many other pretty blooms.

RHODODENDRONS AND AZALEAS

The rhododendrons are a large and ancient genus of flowering bushes belonging to the heather family Ericaceae. There are a few species in North America and Europe, and quite a lot on the mountains of the East Indies and New Guinea, but about 400 species are distributed in the Himalayan and Chinese region.

Most rhododendrons are quite large, forming trees and shrubs with rather leathery leaves, often with downy undersides and tips sheathed in scaly buds through the winter. Their splendid bunches of flowers are terminal, and often open before the leaves, so that in spring and early summer bushes give a mass display of colour over their whole surface.

Azaleas are small rather delicate rhododendrons. They differ from the larger species in their habit of losing their leaves in winter, rather than keeping them for several years, but this is not a strict botanical distinction and azaleas are all included within the genus.

In cultivation, rhododendrons are readily hybridized and as a result there are many hundreds of hybrid forms found in gardens and collections. Most of these stem from a period of considerable rhododendron mania during the late nineteenth and early twentieth centuries, and the major collecting expeditions of E.H. Wilson, J.F. Rock, F. Kingdon Ward and George Forrest. Because of increasing human development in China's remote areas, some of the species gathered by Wilson are now facing extinction in their homeland. Through a project organized by British botanists, rare rhododendrons will be returned to Chinese botanical gardens, where seeds and cuttings can be taken to re-establish wild populations.

There are few sights in the Himalayan forests as impressive as the 10-metre (35-foot) tall tree rhododendrons (*Rhododendron arboretum*) when these are in scarlet flower. Other attractive species are pink, mauve, white and bluish. Most can be eaten by cattle and other ungulates but several are poisonous and liable to be cut down by herdsmen.

Evergreen rhododendrons do well in subalpine forests, often forming a distinct undergrowth with dwarf bamboos and, indeed, dominating the vegetation more and more towards the treeline as the forests become thinner and the trees more stunted. Above the treeline there is often a scrub zone made up largely of rhododendron and juniper. Even on the highly exposed alpine meadows we still find dwarf rhododendrons, such as the pretty *R. nivale* which is the dominant shrub over a large area of the Himalayas.

As well as being beautiful, and prized in cultivation, rhododendrons serve an important role in the forest ecology. Their pioneering ability is a prime factor in allowing tree colonization on poor soils at high altitudes. Their hard roots help to consolidate the mossy, peaty mountain soil, reducing erosion and increasing water penetration. Also, their flowers are enormously important to the insect and bird fauna of these forests. Rhododendrons produce a rich nectar which attracts bees, flies, moths, butterflies and nectar-feeding birds. In addition, many insectivorous birds such as warblers feast around the flowering bushes, picking off the attracted flies. The flowering period almost seems to be planned, with different rhododendrons gradually coming into bloom at increasingly higher elevations, exactly in time with the seasonal altitudinal migration of many mountain bird species, providing vital foods in the spring period before the open alpine meadow flowers burst into life later in the summer.

PEONIES

The peony has been the national flower of China for over a thousand years, celebrated in both paintings and poetry. Peony-patterned celadon bowls of the Song dynasty are among the highest forms of art ever produced in China.

The wild montane *Paeonia suffruticosa* grows in the mountains of Shaanxi, Gansu and Sichuan, in exactly the distribution area of the Giant Panda, but was discovered to science much earlier than that elusive creature. The species had already been introduced into gardens of south-east China as early as AD 750 and created an unrivalled floral fashion.

The Emperor Tang Ming Huang ordered the planting of 10,000 bushes in the grounds of his summer palace. By the turn of the millennium some 90 different cultivars had been developed and methods for grafting and cultivation were well established. It took another 700 years for the peony to be introduced to Europe but it is now a very familiar plant in gardens of all temperate regions.

The Shaoyao *P. lactiflora* has an even longer history of cultivation and was mentioned in records as early as 900 BC. It became most popular in the Hangzhou region, and in the gardens of the famous Chu family there were said to have been no fewer than 60,000 plants of the species.

This ancient science of peony cultivation is all the more remarkable as they are not particularly easy plants to grow – the germination period for their seeds may be as long as two years. Flower buds which are sheltered by forest trees in the wild are very susceptible to spring frosts in an open garden and the plants demand a humus-rich soil in which to grow.

CHRYSANTHEMUMS

Another flower group to attract early horticultural attention in China were the chrysanthemums. Chinese textbooks dating back to 500 BC mention yellow-flowering chrysanthemums, probably the wild *Chrysanthemum indicum*, but by the Tang dynasty there were already white and purple species and, like the peonies, chrysanthemum flower patterns started to appear on stoneware ceramics. By the thirteenth century China had produced many new cultivars and introduced them to Japan and Korea. It is sometimes difficult to remember, when admiring the amazing colours and pom-pom varieties of modern chrysanthemums at a flower show, that these are the products of centuries of cultivation by long-vanished plant enthusiasts.

The Chinese believe the dew that forms on the petals of the chrysanthemum helps restore vitality. Some practitioners advocate eating petals regularly and many medicines are brewed from the dried flowerheads. In southern China *juhua cha* or chrysanthemum tea is a standard on offer at morning *dimsum* restaurants.

CONIFERS AND GYMNOSPERMS

China contains no less than 40 per cent of the world total of gymnosperms – a class of primitive plants distinguished from higher flowering plants by having naked seeds without an ovule. They include the primitive cycads, which are fern-like trees, and also the segmented lianas called *Gnetum*. However, most of China's gymnosperms are the more familiar conifers that dominate the subalpine forest formations of the montane regions.

Perhaps the best-known conifers are the pines (*Pinus*) with their conical cones and long, evergreen needle-like leaves. In

ABOVE LEFT Rhododendrons, such as this beautiful *Rhododendron campanulatum*, have attracted more worldwide interest than almost any other group in China's rich flora and there are societies devoted to their promotion in many countries.

ABOVE RIGHT The scarlet trumpets of *R. planetum* are typical of the startling flamboyance of the genus.

RIGHT The wild montane peony *Paeonia veitchii* was discovered by Chinese gardeners almost a thousand years ago.

BELOW *P. delaveyi lutea* is one of the many magnificent peonies brought into cultivation in China.

China, pines are found as tropical and subtropical formations at moderate altitudes on steep ridges or on fast-draining sandy soils. In the north they are a dominant forest type, again on rather sandy soils. In all there are more than 30 species of pine, from the dwarf *Pinus pumila* in the taiga forest of the north-east to the stately Yunnan pines (*P. merkusii*) of the south-west. Pines are also used extensively in plantation forestry and two species have proved most useful in this respect – the Northern Pine (*P. sylvestris*) in the temperate regions and the endemic Chinese Pine (*P. massoniana*) of the south-east in the subtropical zone.

However, the great subalpine conifer forests of China are not dominated by pines, though some do occur, but they are composed of three main genera – silver firs (*Abies*), hemlocks (*Tsuga*) and spruces (*Picea*). All grow to very great size and have their needles distributed evenly and densely along the twigs. There are several species but, generally, silver firs have tall erect cones, which are blue when young, hemlocks have small, rounded smooth-scaled cones and spruces have long pendulous cones. Firs often dominate the highest forests with almost pure stands at the treeline but mix in with the other two genera at lower elevations. Hemlocks are often the largest trees in the forest and some wildlife species, such as bears and Giant Pandas, are rather dependent on finding old hollow-stemmed hemlocks for their breeding dens. The only resemblance between hemlocks and the poisonous umbelliferous herbs of the same name is the aroma of their crushed foliage.

All the above genera are evergreen, but one common genus of Chinese conifers, the larches (*Larix*), is deciduous. New larch leaves are a glorious bright green, maturing in summer to a

ABOVE The Ginkgo (*Ginkgo biloba*) is a tree commonly seen planted in the grounds of Chinese temples. In prehistoric times it was found all over the world, but was long thought to be extinct until discovered surviving in China. BELOW The Ginkgo leaf is revered because its two lobes are believed to symbolize the principle of perfect balance, or *Yin* and *Yang*.

ABOVE Larches (*Larix*) are deciduous members of the conifer family. Their leaves are a light, bright green in spring, gradually darkening and then turning a glorious golden yellow before falling in the autumn.

darker and more greyish green but changing again to a beautiful golden yellow in the autumn prior to leaf fall. The cones of larch are small and upright, opening whilst still on the branches, and the needles grow in whorls from woody nodes at intervals along the twigs. In the north-east of China, great areas of forest are dominated by larches but elsewhere these appear as occasional trees mixed in with other conifers, lining streams or growing individually in small clearings where they can get plenty of light.

Junipers are another group of conifers well represented in the

forests of China. They are particularly well adapted to the extreme climate on and above the treeline, and often form a mixed shrub layer of dwarf trees mixed with rhododendron shrubs at the interface between forest and alpine meadow. Junipers also thrive in some of the rather barren and fast-draining limestone country in the central regions.

There are numerous other conifer genera present in China represented by only a single species, or up to three species. In the high mountains of the tropical zone are found such interesting species as the fragrant-wooded *Fokienia hodginsii* and the magnificent huge *Keteleeria davidii*. In the Himalayan mountains ancient *Cupressus* trees grow near Tibetan temples. These can reach enormous size. A specimen in the Nyanghe valley measures almost 6 metres (20 feet) in diameter and is estimated to be over 2,000 years old.

Most interesting are some of the relict species of genera that were formerly more speciated and widespread. Examples include the *Taiwania* which now has a disjunctive distribution in Taiwan, Yunnan and the upper Yangtze region.

A notable member of the *Sequoia* family is the Chinese Fir (*Cunninghamia lanceolata*) which grows wild in south-east China but is now an enormously important forestry plantation species. The leaves of *Cunninghamia* are flattened, hard and extremely sharp-spined.

Gymnosperms include the yews (*Taxus*). These trees have single ovules that are not on a scale, as in true conifers, but surrounded when ripe by a fleshy colourful aril which attracts birds that act as distributing agents for the seeds. Yew trees are generally found in the upper zones of mixed conifer forests but are rarely dominant. They are placed in their own order, Taxoles.

Another Chinese gymnosperm enjoys an entire order to itself – namely the famous Maidenhair Tree or Ginkgo. *Ginkgo biloba* grows spontaneously in western Tianmushan in hilly country of the lower Yangtze valley, appearing in mixed stands of broadleaf forest. It also appears as a wild plant on the remote hills along the Anhui-Zhejiang border and in eastern Guizhou. This scattered distribution suggests that it was once more widespread and may have been eliminated from natural forests by centuries of selective logging of the best trees by Chinese farmers.

Today the Ginkgo is much more numerous as a cultivated tree. It is frequently planted around sacred sites, and single trees are themselves often treated as sacred by local people who decorate them with paint or garlands, and burn joss sticks and mumble prayers at their base. Ginkgos have a lovely growth form and attractively shaped two-lobed leaves which turn a rich golden yellow in the autumn before falling. These leaves are considered lucky by the Chinese, as the two lobes are believed to represent perfect balance. The plant's Chinese name is *yin xing* or 'silver fruit', the monetary connotation being another reason to regard it as lucky.

The Ginkgo is a true 'living fossil', almost unchanged for 200 million years, and the last relict species of a group that was once common and widespread over much of the world. It is as well that its 'specialness' was recognized by ancient Chinese scholars and that it has been so well preserved in temple gardens.

EASTERN PLANTS IN WESTERN GARDENS

In much of the West, the well-known and well-loved plants found in many a back garden are to a very large extent Chinese. This is due largely to the extraordinary work of the collectors who, at the turn of the century, brought back the seeds of a huge number of splendid varieties, particularly into Europe.

It is a remarkable experience to take a country walk through the forests and scrublands of Sichuan and recognize so many of the native plants that grow on every side. Great masses of purple-flowering *Buddleia davidii* flourish on wasteland and in thickets along the gravelly river banks, their tubular florets smothered in Painted Lady, Tortoiseshell and gaudy Peacock butterflies. Between the trees the ground is carpeted with the familiar simple flowers of anemones, another common garden species. Where the ground is more rocky, the tight-packed flowers of *Cotoneaster horizontalis* hug the flattened branches that are so ideal for training against the side of houses. Just behind, the yellow flowers of a small bush look familiar, and the thorny stems and little holly-like leaves identify it as the Barberry (*Berberis vulgaris*). Climbing over the shrubs is another pretty garden favourite, *Clematis armandii*, and beside it is a flowering bush of *Hydrangea aspera*.

Western garden shrubberies abound in Chinese bushes and evergreens. Many species of rhododendrons have been reared and crossed to fill the nursery catalogues. China is also famous as the home of the camellias, a group of very attractive evergreen shrubs with splendid, well-formed flowers, much favoured by gardeners wherever winter conditions are mild enough for their growth. Other ornamental shrubs include Spineless Holly (*Ilex pedunculosa*), the scarlet-fruited Firethorn (*Pyracantha coccinea*) and some varieties of the privet *Ligustrum ovalifolium*. Conifers include Dawn Redwood, Chinese Fir and Golden Larch, while among the broadleaved species are the delicate white-fruited Hupeh Rowan (*Sorbus hupehensis*) and flowering Japanese Cherry (*Prunus serrulata*).

The purple-flowering *Buddleia davidii* is native to Sichuan province but has become a familiar shrub in Western gardens, where it is known as the butterfly bush. The Marbled White butterfly (*Melanargia montana*) seen here is not unlike its European counterpart.

Several of the local maples are also common in Western gardens. Commonest of all is the Japanese Maple (*Acer palmatum*) which has deep, sharply toothed lobes to the delicate leaves that turn a deep scarlet in autumn. Other favourites are *A. henryi* with its three separate leaflets and the Paperbark Maple (*A. griseum*) with its flaking papery bark. *A. davidii* has a simple ovate leaf but the little two-winged fruits are recognizably maple. Less common, but equally appreciated, Chinese trees include the Locust Tree (*Robinia* sp.), Pagoda Tree (*Sophora japonica*), Chinese Tulip Tree (*Liriodendron chinense*), Dove-tree (*Davidia involucrata*), the trumpet-flowered *Paulownia tomentosa*, Nyssa (*Nyssa sinensis*) and Chinese Fringe Tree (*Chionanthus retusus*).

The tropical evergreen forests of China are characterized by a high density of epiphytic orchids and ferns. The Bird's-nest Fern (*Asplenium nidum-avis*) is one of the largest and most splendid of these epiphytes.

A decorative camellia (*Camellia reticulata* hybrid) in Yunnan. Hundreds of attractive horticultural varieties have been developed from the wild species, which is closely related to tea.

Dwarf Quince (*Chaenomeles speciosa*) is another ornamental bush that grows wild in central China. In early summer it is decorated by pretty, white to red, five-petalled flowers and these mature into golden fruits that look like apples. And what garden would be the same without the bright yellow flowers of the *Forsythia* bushes that also made the long journey to Europe from central China. Lilacs (*Syringia*), *Daphne* species and the red-berried *Viburnum betulifolium* all made the same trip.

The beautiful *Magnolia officinalis* flowers after the leaves have formed. The flowers are creamy-white and fragrant with the petals arranged in a dish shape, and the stamens have long red filaments. The flowers of *M. dawsoniana* are much pinker and held horizontally, generally opening before the leaves have developed in early spring. *M. delavayi* has just six, broad, spoon-shaped petals of a rich buttery colour. But most splendid of all is the glorious *M. campbellii*, which flowers on the bare branches of late winter. The flowers are huge and pink with the outer petals opened whilst the central ones remain half closed to give a curious cup-and-saucer appearance. The stark contrast of these big pink blooms against the dark green and often misty forest canopy of late winter is a truly splendid sight.

Another bush which flowers before the leaves form is the *Bauhinia variabilis*. These trees grow in secondary forests, often overhanging the roads that wind through mountainous countryside, and can give a whole valley a festive appearance in early spring. Several species of *Bauhinia* are favoured for cultivation and one species, *B. blakeana*, is the official flower of Hong Kong.

Chrysanthemums and peonies are, of course, Chinese classics and many other flowers of annual and perennial beds, such as geraniums, delphiniums, irises, begonias, lilies and poppies, were also brought back from China. So, too, were the alpines beloved as rockery plants – asters, primulas and polygonums.

Nor are imports to Western gardens limited to the ornamentals. Many fruits and vegetables are also found wild in the Chinese countryside. Peaches, redcurrants, blackcurrants, gooseberries, cherries, pears, plums, strawberries, rhubarb, grapefruits, kiwifruits and other everyday fare either originated in China or have been hybridized with wild Chinese varieties.

China's Native Fauna

China's native fauna consists of a mixture of tundra species from northern Europe, many widespread genera of temperate creatures, the specially adapted wildlife of the high plateau lands, most of the Himalayan species of the very high mountains and a largely unique subtropical fauna in the central and south-east regions. Added to all of these is the tropical fauna that is prevalent in south-east Tibet, south Yunnan, extreme south Guangxi and Guangdong, south Taiwan and Hainan.

MAMMALS

Ungulates include a wide range of deer types from the Reindeer (*Rangifer tarandus*) and Moose (*Cervus alces*) of the most northerly latitudes to the many deer of the temperate zone, and the dainty mousedeer (*Tragulus*) and muntjacs (*Muntiacus*) of the south. Some endemic species include the Chinese Water Deer (*Hydropotes inermis*) of the eastern lowlands, the White-lipped Deer (*Cervus albirostris*) of the Tibetan plateau, the Tufted Deer (*Elaphodus cephalophus*) of the central mountain ranges and two endemic muntjac or barking deer (*Muntiacus reevesii* and *M. crinifrons*). The Brow-antlered Deer (*Cervus eldi*) is extinct in China, except for a captive herd maintained on Hainan, and the endemic Père David's Deer (*Elaphurus davidianus*) did become extinct in the wild but is now being reintroduced from European zoo stock. Muskdeer (*Moschus*) are small mountain-living deer with stiff bristly hair and large canines. They lack antlers, but the males have a pod or gland of musk oil which is highly prized in the making of many Chinese medicines and, as a result, the unfortunate creatures are faced by constant rows of traplines throughout their habitat.

The steppes and high plateaux support some gazelles and antelopes, and several wild sheep, but the strangest ovids are the group known as goat-antelopes which include the shaggy black Serow (*Capricornis sumatraensis*), the delicate brown Goral (*Nemorhaedus goral*) and the large, sad-looking Takin (*Budorcas taxicolor*). True cattle include the Wild Yaks (*Bos grunniens*) of the high plateaux and the great Gaur (*B. gaurus*) of the tropical forests in the south-west.

The Bactrian Camel (*Camelus bactrianus*) still lives in the semi-desert regions of east Xinjiang and west Gansu, ranging across to the Mongolian Gobi desert. This is the wild ancestor of the domestic two-humped camel used by traders and farmers over much of northern China and central Asia. Populations in the wild have dropped to a few hundred animals as a result of hunting, disturbance from mining activities and climatic warming. That camels survive at all is due to their extraordinary hardiness, and many live for over 40 years. They can exist not only in areas of extreme heat and aridity, where they are able to go as long as 20 days without food or water, but are also well adapted to live at extreme altitudes in the cold snowy weather of the plateau and mountain areas.

Other ungulates include the wild horses. Przewalski's Horse (*Equus ferus*) is extinct in the wild, though is in the process of being reintroduced from captive stock, but the Wild Asses (*Equus hemionus*) of the deserts and plateaux can still be found in some numbers. Indian and Javan Rhinoceroses formerly occurred in south-west China but have not been reported for many years.

The ubiquitous Wild Pig (*Sus scrofa*) ranges in almost all habitat types – a single species, varying from a huge hairy beast in the extreme north to a smaller, half-naked pig of the tropical forests in the south. Asian Elephants (*Elephas maximus*) survive only in the south of Yunnan province.

Rodent herbivores are very diverse, with squirrels, rats,

Four races of Tiger (*Panthera tigris*) occur in China. Although these animals are a powerful symbol in Chinese mythology and art, all are endangered by the belief that their bones are a potent medicine. In some regions, tiger breeding farms were set up as sources of supply but since 1993 such trade has been banned and the farms have become bankrupt.

A rare cub of the large Siberian race (*P.t. altaica*) looks like an enchanting pet but will soon grow into a fearless killing machine, weighing 250 kilograms (550 pounds) and able to tackle the largest ungulates in the forest. No more than 10 Siberian Tigers remain in the wild in China.

bamboo rats and porcupines in the southern forests, voles, picas and marmots on the plateaux and long-legged gerbils in the northern deserts.

Of the carnivores, a few such as the two panda species and the Black Bear (*Ursus thibetanus*) are in fact almost totally vegetarian. Others, such as civets, badgers, mongooses and Brown Bear (*U. arctos*), can be classed as omnivores, while the many species of weasels, cats and dogs are truly predatory. Several otters (*Lutra*) are specialized for catching fish.

Only a handful of Siberian Tigers (*Panthera tigris altaica*) remain in China, with a few more in Russia and North Korea. These are the largest race of tiger with adult males reaching over 250 kilograms (550 pounds) in weight. They can kill deer and young moose but their main prey is wild pigs. The Chinese government has set up a captive breeding programme for the species and keeps 70 animals on a farm in Heilongjiang. There are plans to release some of these huge carnivores back into the wild, but whether there are large enough blocks of forest for these animals to roam in without becoming a pest is rather questionable. The cubs are beautiful and cuddly but soon grow into dangerous killing machines.

Other races of tiger in China include the Bengal Tiger, that occasionally wanders into Chinese territory of south-east Tibet, the Corbett or Indo-China Tiger that has all but vanished from south Yunnan, and the South China Tiger which still survives in a few isolated rugged mountain areas of southern China.

The Black Bear is much more common and widespread in China than the Giant Panda. It will raid birds' nests and bee hives, as well as preying on small animals, but its main diet is fruit and other vegetable materials. In autumn bears stock up on acorns and beech nuts. When eating acorns they often pull all the outer branches in towards the centre of the tree crown, making a kind of nest to sit in but leaving a tell-tale broken mess in the canopy. They also roll in brambles to pull the straggling plants together and make collecting the fruit easier, and are serious pests in apple and walnut orchards. In winter, unlike pandas, they hibernate in sheltered dens in rock crevices, caves or hollow trees, emerging thin and hungry in the spring to have their cubs. Bears often have three or four cubs at a time, so can breed much faster than the Giant Panda.

Bear paws and other body parts have long been held in esteem as a medicinal food in China whilst the crystals of gall from their gall bladders are a valuable ingredient in many forms of traditional Chinese medicine. For this reason many bears were killed each year and the population was seriously threatened. Since joining CITES (Convention on International Trade in Endangered Species) China has banned all hunting or trade in wild bears. However, it is estimated that some 8,000 bears have been brought into farms, where they are still reared for tapping of the gall bile to make medicine. Considerable international controversy remains as to the ethics of continuing this practice but if it is decided to close down such farms there is also the problem of what to do with 8,000 human-habituated but dangerous animals.

There are four species of gibbons in China, all confined to the extreme southern fringe. The White-browed Gibbon (*Hylobates hoolock*) is found in south-east Tibet and in south Yunnan to the west of the Nujiang river. The White-handed Gibbon (*H. lar*) occurs only in a tiny area between the Nujiang and Lancangjiang rivers, though its range extends south of this right down as far as northern Sumatra, and there are historical indications that this species may formerly have extended further north in China. It has a buff to brown fluffy coat, a white ring round its bare black face and smart white hands and feet.

The Black or Concolor Gibbon (*H. concolor*) is all black and lives in south-west Hainan and in the Ailaoshan and Wuliangshan mountains of southern Yunnan. The White-cheeked Gibbon (*H. leucogenys*) is found only in the extreme south of Yunnan in Xishuangbanna to the east of the Lancangjiang river and is also very restricted in distribution.

All these gibbon forms are now extremely rare, in part because they are very easy to hunt. They can be attracted to imitations of their own calls and once one animal has been shot it can be tied up in a tree where it attracts the rest of the family to be shot one by one. In addition, gibbons are highly territorial animals with low density in the forest. Families rarely number more than four or five animals and each family patrols a territory of 2 to 5 square kilometres (¾–2 square miles).

Gibbons are frugivores, with a particular liking for strangling figs, but they also eat other fruits, some insects and plenty of young leaves and fresh shoots. Early in the morning they give wonderful haunting group choruses which vary from species to species but are all far-carrying long-drawn-out calls of rising and falling pitch.

Other Chinese primates include several macaque species (*Macaca*), which live in large groups and travel largely on the ground, feeding on fruits and shoots. In addition, there are some elegant leaf monkeys (*Trachypithecus*) that are mostly arboreal and feed on foliage. The rare black-headed Francois' Leaf Monkey (*T. francoisi*) and the closely-related White-headed Leaf Monkey (*T. leucocephalus*) are endemic to the limestone forests of Guangxi province, where they lead a precarious existence bouncing about on the cliffs and hanging vegetation in the rugged karst hills. The babies are a rich golden colour in stark contrast to their black parents. Angry patriarchs give loud nasal calls. These two leaf monkeys were formerly thought to be subspecies of one single species but, as they live side by side on the same limestone blocks, this now looks unlikely and they have been given full specific status.

Perhaps the most bizarre of China's monkeys are the three species of snub-nosed monkey (*Rhinopithecus*). These huge shaggy-coated monkeys live in very large troops of over a hundred individuals. They inhabit the subalpine forests of the central regions where they eat the cabbage lichens growing on huge trees or strip the young leaves from birch and *Sorbus* trees. The commonest species is the Golden Monkey (*R. roxellanae*) of Sichuan, Shaanxi and Hubei provinces. The entire fur is golden yellow but the face is a strange blue colour, decorated by a small upturned nose and wattles at the side of the mouth. A rarer form, the Grey Snub-nosed Monkey (*R. brelichii*), lives in Guizhou and is now confined to Fanjing mountain, whilst a less well-known species, Biet's Snub-nosed Monkey (*R. bietii*) lives in the north-west corner of Yunnan and the south-east corner of Tibet.

The last primate is the small nocturnal Slow Loris (*Nycticebus coucang*) or, as it is known in Chinese, *lanhou* or 'lazy monkey'. This is a common animal in secondary forest, and occasionally in primary forest, where it feeds on fruits, insects and leaf shoots. Lorises are easily seen at night in torchlight and hunters collect them as pets or to use their body parts as medicine. As a result there are fewer of them than there ought to be.

Pandas. There are only two species of panda in the world, the famous Giant Panda (*Ailuropoda melanoleuca*) and its less well known but really beautiful little cousin the Lesser or Red Panda (*Ailurus fulgens*), a racoon-like animal with bright reddish fur and a long banded tail. Both occur in China, the Giant Panda exclusively, but the Red also extends along the south face of the Himalayas as far as west Nepal.

Female Giant Pandas make dens in hollow trees or rock clefts where they have their babies. Usually two cubs are born but only one survives the first few weeks. This is the most strenuous survival test in the animal's life. Once weaned, a Giant Panda faces few enemies. Large enough to fend off bears, leopards and other potential predators, the species has a generally low mortality rate and, with such low reproductive rates, populations can grow gradually.

The Red Panda is largely arboreal and eats fruits and leaves in the treetops. The Giant Panda also sometimes climbs trees to play, sleep in the warm rays of the sun or to escape predators or more dominant pandas on the ground. Otherwise it is terrestrial and walks with a curious pigeon-toed gait. Both species urine mark their territories and the smell lingers on the still forest air for many hours.

Pandas feed extensively on bamboos and have special adaptations to their dentition, gut and forepaws to accommodate this diet. Bamboo is a woody grass, but so tough are its stems and so silicaceous its leaves that it is difficult to eat and even more difficult to digest. Both pandas leave copious firm pellets of only partly digested bamboo and both have a curiously formed opposable thumb which enables them to hold bamboo stems firmly whilst stripping them of leaves.

Every 30 years or so, bamboos flower gregariously over large areas and, after flowering and seeding, the mature plants die. The Giant Panda populations, in particular, are jolted by sudden food shortage. Some animals starve, some survive on the reduced supply of bamboos that fail to flower and others migrate out of their valleys in search of new areas where the bamboos at lower

ABOVE The Giant Panda (*Ailuropoda melanoleuca*) is China's best-known and most-loved animal. It has become a symbol of conservation around the world as the logo of the World Wide Fund For Nature.

BELOW The Giant Panda's small cousin is the less well-known Red Panda (*Ailurus fulgens*). This is a pretty racoon-like creature with a long banded tail. It feeds mostly on bamboo, as does its larger relative, but also eats some fruit.

The Red Panda is hunted for its attractive fur and is now nearly as endangered as the Giant Panda. Fortunately, the animals breed well in captivity and have a fairly wide distribution, extending to east Nepal.

altitudes or in further valleys have not yet flowered. In former times these flowering incidents probably caused the pandas only a moderate tightening of the belt, and provided an important incentive for migration and hence cross-breeding between populations that would otherwise tend towards inbreeding.

Today, however, these flowering events can be very serious because farmers have encroached into all the lower valleys of the panda habitat, and deforestation within the Giant Panda's range has caused severe fragmentation of the overall population. There are no longer any bamboos at lower altitudes and pandas wandering to find new valleys must cross village lands. WWF and the Ministry of Forestry have prepared an ambitious masterplan which has called for a halt to logging and forest clearance in the range of the Giant Panda, an increase in the number and area of nature reserves, stricter controls on hunting in the animal's range and the maintenance, and even restoration, of the habitat corridors that connect adjacent fragmenting populations.

Saving the Giant Panda is a major challenge for China and the animal has become a recognized symbol of Chinese wildlife and conservation efforts. There are still over a thousand of these beautiful and lovable creatures in the crescent of mountain ranges that run from west Sichuan across to south Gansu and south Shaanxi provinces. Saving the species is quite feasible and in protecting its habitat conservationists are also saving the habitat of an enormous range of other important species.

Sheep and Goats. China has a total of 10 species of goats and sheep – more than any other country. Wild goats include the serows and three species of goral. Serow are large and shaggy with long manes, big ears and recurved, tightly annulated horns. The Himalayan Serow (*Capricornis sumatraensis*) has a wide distribution along the Himalayas, across most of central China and then south down to Indonesia, while the Taiwan Serow (*C. formosae*) is confined to Taiwan. Both live on rocky and steep terrain, resting in caves and crevices, and are generally solitary.

The smaller goral has much more delicate horns, is paler in colour and lacks a mane, though it often has a whitish throat-collar. The Himalayan Goral (*Nemorhaedus goral*) is replaced over central China by the Chinese Goral (*N. caudatus*), and the Red Goral (*N. baileyi*) is found only along the China-Myanmar border in the Gaoligong mountains. Gorals live in pairs or small parties, and feed on grasses. Another goat, the Himalayan Tahr (*Hemitragus jemlahicus*), occurs on the southern faces of the Himalayas and lives in flocks. It may occur in the Chumbi valley and a few other marginal parts of southern Tibet.

Largest and strangest-looking of the goat-relatives is the Takin (*Budorcas taxicolor*), distributed from Bhutan through the east Himalayas, west Yunnan and Sichuan through the Qinling mountains of Shaanxi. Takin have a big muzzle like a moose and dangerous horns like an African wildebeest. They live in herds of varying numbers and feed on alpine meadows and forest undergrowth, making regular visits to salt licks.

Commonest of the sheep in China is the Blue Sheep or Bharal (*Pseudois nayaur*) which is widely distributed along the Himalayas, across the eastern end of the Tibetan plateau and into the mountains of west Sichuan and Yunnan. Bharal live in large herds, grazing over the alpine grasslands. Much larger are the

ABOVE Blue Sheep (*Pseudois nayaur*) are the commonest and most widespread of the wild sheep in China. They occur across the whole Tibetan plateau, feeding in large herds of a hundred or more on the alpine meadows of the highest mountains. RIGHT The Takin (*Budorcas taxicolor*) is a large shaggy-coated goat-antelope. These animals live in herds, feeding on open alpine pastures in summer but descending into the conifer and mixed forests in winter.

Marco Polo Sheep or Argali (*Ovis ammon*) which have heavy recurved and strongly annulated horns reminiscent of fossil ammonites. Several races occur along the northern faces of the Himalayas but all are highly prized by hunters and are now very rare. Even more rare is the Markhor (*Capra falconeri*), a wild goat with long twisted horns, that may still occur in the Taxagorgan area of Xinjiang on the Pakistan-Afghanistan border. The splendid Ibex (*Capra ibex*), with its swept-back curved horns, lives in the high mountains of north-west China. The males are decorated by long beards. They live in large herds feeding on the sparse grasses. Wherever the wild sheep and goats wander, the Snow Leopard stalks, but their real enemy is man and the telescopic sights of high-velocity rifles.

OTHER GROUPS

Riches in other faunal groups include a wealth of butterflies and other insects, especially in the tropical regions, and many endemic species of amphibians and reptiles. One endemic group of interest are the giant salamanders. Several species live in the upper rivers of the Yangtze system but the largest of all, *Megalobatrachus davidianus*, known in China as 'fish baby', can reach a weight of several kilograms and is found in both the Yangtze and Yellow river systems.

BELOW LEFT The huge Atlas Moth (*Archaeoattacus edwardsii*) is quite common in the tropical forests of southern China. The pupa is densely cocooned in silk. BELOW RIGHT The beautiful Red Lacewing butterfly (*Cethosia biblis*) is a similarly common insect in southern woodlands.

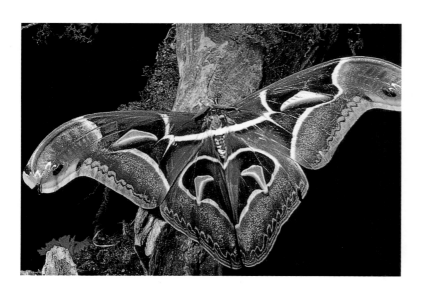

BIRDS

The wealth and diversity of China's bird fauna reflects the country's wide range of habitat types and associated climatic variations. Most temperate bird families are well represented and in the tropical southern fringe are found good numbers of the tropical Asiatic families, such as loud honking hornbills, pretty broadbills, leafbirds, colourful sunbirds, flowerpeckers and other forest species.

China is the most important country in Asia for wetland birds. Species include cranes, storks, ibises, herons, ducks and swans, rails, grebes and waders. Quite a number of rather spectacular wetland birds are nearly endemic to China, such as the splendid Mandarin Duck (*Aix galericulata*), Scaly-sided Merganser (*Mergus squamatus*), Baikal Teal (*Anas formosa*) and Falcated Duck (*A. falcata*).

Other significant families are the resplendent pheasants for which China is a major distribution centre; the birds of prey that are particularly prominent over the wide spaces of the extensive plateaux, deserts and grasslands; and high-altitude specialists in the Himalayas, Tibetan plateau and other high mountain ranges of the country. The central region around the Sichuan basin is a particular hotspot for endemic forms of many families – laughingthrushes, warblers and others.

Many of China's birds are migratory. Typically, these species breed in the northern summer at high latitudes and migrate in winter to the south of the country. A few species breed beyond China in the Arctic Circle and are only winter visitors or passage migrants.

Pheasants. China is the global centre of distribution of the pheasant family Phasianidae, supporting 52 species out of a world total of 196. Pheasants are among the most spectacular-looking birds, being generally large and with many species exhibiting brilliant coloration of plumage and extravagant ornamentation of plumes, ear-tufts, ruffs, brightly coloured bare facial skin, inflatable pouches, or iridescent ocelli on their feathers.

The family is highly diversified in its use of habitat. Many of the species are tropical, including the splendid Green Peafowl (*Pavo muticus*) whose males have a huge tail which can be fanned out and shaken in display to reveal a stunning array of colourful eyespots. The smaller but equally dazzling Grey Peacock Pheasant (*Polyplectron bicalcaratum*) also lives in tropical hill forests. Many more species live in temperate forests and scrub, including the familiar Common Pheasant (*Phasianus colchicus*) which, in addition to being beautifully marked, is also the most popular and widespread gamebird as a result of its introduction to other countries. Some of the other temperate pheasants of China are even more sensational. The plumage of the Golden Pheasant (*Chrysolophus pictus*) is a stunning combination of red, gold, green and metallic blue, with a bright yellow silky crest and rump and a gorgeous mane of golden fanlike feathers edged smartly in black.

The Lady Amherst's Pheasant (*C. amherstiae*) is yet more magnificent, with an enormously elongated black-and-white barred tail trimmed with scarlet flag-like basal feathers. The nape is a white mane barred with black edges and set off by a scarlet crest and metallic green crown and breast. In terms of tail length, the Reeves's Pheasant (*Syrmaticus reevesii*) outdoes them all. This bird's banded tail feathers are 1.5 metres (5 feet) long and are traditionally used as head-dress adornments in the famous classical Beijing Opera.

In the subalpine zone, gorgeously bronzed Monal pheasants live among the rhododendrons with families of short-tailed tragopans. Different species of tragopans are found in different parts of China, but all are remarkably colourful with bright spotted plumage and inflatable throat pouches.

Even in the tundra, desert and harshest parts of the high plateau we still find pheasants. Tibetan Snowcock (*Tetraogallus tibetanus*) and Snow Partridge (*Lerwa lerwa*) live in the icy cold of the Tibetan plateau. Himalayan Snowcocks (*Tetraogallus himalayensis*) range up to 6,000 metres (20,000 feet) on some of the highest mountains in the world. Chukars (*Alectoris chukar*) live in some of the most barren deserts of China.

Perhaps the most important pheasant of all is the Red Junglefowl (*Gallus gallus*), the wild ancestor of the domestic chicken and now the commonest bird on earth. Junglefowl are frequently seen and heard in the tropical scrub forests of southern China. The cocks crow each morning with a call much like that of the familiar barnyard bird, but with a slightly truncated crow and shriller tone.

All the pheasants feed on the ground, eating seeds, insects, fruits and fresh leaves. Most live in small parties in dense thickets where they can scuttle quickly out of harm's way if disturbed by predators. Nests are usually made on the ground, though most species roost at night in the safety of trees. Only the quails and some partridges curl down tight in the open land to spend a cold night under the stars.

Cranes. Cranes are another family of birds for which China is rightly famed. Out of a world total of only 15 species, China has no fewer than eight. Cranes are splendid, tall, proud-looking birds and also beautifully decorated. They live on the open plains, steppe grasslands, semi-deserts and in marshy areas. Superficially, cranes look like storks or herons but their bills are much shorter and their legs and necks are proportionally longer, being fully extended in flight. They feed on vegetable materials, including roots and tubers of small swamp plants and grasses, seeds, grain, and also some invertebrates or even small vertebrates when they can catch them.

One feature of crane behaviour that has captured the imagination of their human co-inhabitants wherever they occur is their courtship dancing and generally faithful or perfect pair-bonding. In China, cranes are repeatedly honoured in verse, carvings and stylized paintings.

Courting cranes pace around each other with wings half open, periodically leaping up from the ground and then bowing gracefully. All the time they emit excited trumpeting calls. After courtship the huge nest is piled up on the ground in some quiet marshy area and a modest one to three eggs are laid. Emerging young of all species are fluffy sandy orange but the adults of most species are smartly clothed in black and white or black and grey. Some have decorated heads with black, white and red colour combinations.

Largest of China's cranes is the Sarus Crane (*Grus antigone*) which stands 1.5 metres (5 feet) tall and was once a breeding resident in western Yunnan. Sadly, the species is on the decline in South-east Asia and no longer breeds in China.

Most of the other cranes in China are migratory. They breed in the extreme north or up on the high Tibetan plateau but descend to the south and centre of the country in winter. The rare Siberian Crane (*G. leucogeranus*) breeds in Siberia, passes through the great wetlands of north-east China and ends up wintering around the great lakes of the lower Yangtze valley.

The Red-crowned Crane (*G. japonensis*), almost as large as the Sarus, breeds in Heilongjiang and north Jilin and winters in the wetlands of central eastern China. The rare Black-necked Crane (*G. nigricollis*), in contrast, breeds on marshes in Tibet and on the

China supports more than a quarter of the world's pheasant species in a highly diverse range of habitats. The adult males, in particular, are renowned for their brilliant coloration and often ornately embellished plumage. ABOVE The Golden Pheasant (*Chrysolophus pictus*) sports a kaleidoscope of colours with a gorgeous flowing mane.
BELOW LEFT The long white tail feathers of the male Silver Pheasant (*Lophura nycthemera*) are delicately vermiculated with jagged black lines. Regrettably, they are often prized as trophies by hunters.

BELOW RIGHT Swinhoe's Pheasant (*L. swinhoei*) is one of two pheasants endemic to Taiwan. Though highly endangered it is staging a recovery due to the protection now given to its habitat. BOTTOM LEFT The Reeves's Pheasant (*Syrmaticus reevesii*) of central and southern China has an enormously long banded tail. The feathers are traditionally used in head-dresses for the Beijing Opera. BOTTOM RIGHT Lady Amherst's Pheasant (*C. amherstiae*) is one of the most bizarrely decorated. The species, found in the forests of south-west China, is rather rare.

ABOVE The Great Hornbill (*Buceros bicornis*) is very rare in China, being found only in the tropical rainforests of south-west Yunnan, but even so, these birds are too large and noisy to escape the attention of hunters.

BELOW White-naped Cranes (*Grus vipio*) nesting in the Zhalong Crane Reserve. The chicks must grow for several more months before they are strong enough to fly to the Yangtze valley for the winter.

Sichuan-Qinghai border, wintering southwards to south-east Tibet, neighbouring Bhutan, north Yunnan and west Guizhou. Tibetan farmers celebrate the return of the cranes with imitative dances of their courtship, but farmers in Yunnan and Guizhou are somewhat less pleased by the birds' prediliction for eating their potatoes.

Laughingthrushes. Laughingthrushes are another group of birds for which China is the world centre of distribution, with 37 species out of the world total of 53. Laughingthrushes, a subfamily of the babblers, are the size of large thrushes and are often found in flocks, keeping to the lower levels of forest and scrub habitats. They feed mostly on the ground, turning over leaves and scrabbling about for insects and seeds. One or two members of the flock are always on the lookout for danger, keeping to the bushes or trees over their feeding comrades. These lookouts are very acute and give loud alarm calls as soon as they see the slightest suspicious movements.

Flock members make regular contact calls which can erupt into a cacophony of raucous alarm or territorial calls. In some species these are clear, long-carrying whistles but in others they are a riot of maniacal laughter which gives the group its common name. Some species have very beautiful clear songs and are highly prized cage-birds in the Chinese bird trade, more often found hanging in a bamboo cage than flying free in the scrub forests where they are so persecuted by trappers. The Spot-breasted Laughingthrush (*Garrulax merulinus*) is perhaps the most melodious but its distribution in China is confined to Yunnan and it is rather rare. Much more familiar is the Hwamei (*G. canorus*). *Hwamei* in Chinese means 'beautiful eyebrow',

ABOVE The Red-winged Laughingthrush (*Garrulax formosus*) is an endemic species found only in a small area of mountains in the west of Sichuan province.

RIGHT A White-browed Rosefinch (*Carpodacus thura*) among the rhododendron bushes at 4,000 metres (13,120 feet) in the Himalayas.

referring to the white ring around the eye and the clear white browline contrasting with the rich rufous brown of the rest of the bird's plumage. Some town markets will have hundreds of these birds hopping actively within the confines of their cages, singing beautifully to the delight of old Chinese men, who seem particularly drawn to the species.

Some public parks have 'Hwamei corners', where owners bring their treasured pets and hang them up in the trees with those of their friends so that the birds can see one another and sing more wonderfully at the stimulation of being with their own species again. Competitions are sometimes staged to pick the best songster of the morning.

In Chinese *Hwamei*, or just *Mei*, has become the generic name for the whole subfamily which is, in fact, rather varied. Two species – the Red-tailed Laughingthrush (*G. milnei*) and Red-winged Laughingthrush (*G. formosus*) have beautiful crimson tails and wings, whilst the related Red-faced Liocichla (*Liocichla phoenicea*) has bright crimson sides to its head and looks quite splendid. Another striking species is the White-crested Laughingthrush (*G. leucolophus*), which has a crested, pure white head with a broad black eye-stripe and is one of the noisiest 'laughers' of the subfamily, with a wild shrieking group chorus when excited.

Several species live in dense forests but others live in more open areas. The Masked Laughingthrush (*G. perspicillatus*) is one of the commonest birds in parks and gardens of south China and Hong Kong. The Brown-cheeked or Prince Henri's Laughing-thrush (*G. henrici*) lives in scrubby thickets and semi-arid regions of south-west Tibet up to nearly 5,000 metres (16,000 feet) altitude. Various species live in the rhododendron groves of alpine forests in the high Himalayas and right across the regions of central China.

Rosefinches. Any visitor to the alpine scrub and meadows of China will be delighted to see flocks of gorgeous pink, chunky finch-like birds. These are the famous rosefinches but there are 25 different species and they are difficult to identify. All but three North American species of rosefinch occur in China, though many species distribute well outside China into neighbouring countries. The Common Rosefinch (*Carpodacus erythrinus*)

extends its range as far as Europe whilst the Sinai Rosefinch (*C. synoicus*) even extends as far as the Negev and Sinai deserts. However, most rosefinches are centred on the Tibetan plateau or Himalayan region, and variously distributed on the mountains of central and eastern China, with just one species, the Vinaceous Rosefinch (*C. vinaceus*), reaching as far east as Taiwan.

All rosefinches feed on small berries and seeds of trees, bushes, herbs and even grasses. Some peck at fruit or eat rosehips and plant shoots. They have typical finch bills – short, conical and powerful for extracting and crushing small seeds. Males are always pinkish, or crimson with pink. Different species can be distinguished at close range by the distribution of pale brows, foreheads, spots, streaks and wingbars. Females are mostly drab coloured, brown, grey and streaked. In the absence of accompanying males, the females could easily be confused with all manner of other 'little brown birds' but escorted by their gorgeous mates they can be identified correctly. The desert-living Sinai Rosefinch is paler than most, with only a modest pink on the face and throat of the male.

Many of the rosefinches live in areas of extreme seasonal climatic variation. Most accommodate such problems by being altitudinal migrants. In winter, when the alpine zone lies under a deep layer of snow, the rosefinches descend into valleys to inhabit subalpine forests and farmlands. However, as soon as spring arrives and the snows begin to melt they start to move up to higher elevations again, where they will breed among the rhododendrons. By the height of summer they are at maximum altitude, feeding on the seeds of tiny herbs in the open meadows of the alpine zone. Species living at the furthest north of the range of rosefinches are more traditional north-to-south migrants, breeding at the extreme north of their range but moving south into forests and reedbeds for the winter.

The White-browed Rosefinch (*C. thura*) is typical of the group. Like several other species the male has a bright pink eyebrow whilst that of the female is buff, but the extreme rear end of the brow in both sexes is white. This species is quite common and has a wide distribution from Afghanistan and the Himalayas across most of eastern Tibet and north central China. It lives in the rhododendron and juniper scrub close to the treeline but feeds in the alpine meadows in summer and descends into conifer forests for the winter. The call of the species is louder than that of most other rosefinches, with loud whistles and fast, piping bleats. The song is a twittering given from the treetops.

A WALK AMONG THE PANDAS

It is February. Even in Chengdu a rare fall of snow has covered the ground and the drive up the Min valley towards the Wolong Nature Reserve is a journey into deeper winter. The jeep is unheated and draughty, the windscreen wipers creak in a never-ending noisy to-and-fro, and the glimpses of the great muddy Min river are not encouraging.

The journey that usually takes four hours today takes six, but we finally reach the reserve station of Hetaoping. It is here that the Ministry of Forestry have built an ambitious breeding station for the Giant Panda and I have been allocated a small house for the period that I am helping them to prepare a species survival plan to preserve this enigmatic emblem of Chinese conservation. Even in summer the house is cold. Now it is frigid and, outside, huge icicles hang like stalactites. The small stream is partly frozen over and still the snow is falling gently on the graceful swathes of umbrella bamboo.

The station is almost deserted. Everyone who can escape this Siberia-like posting has moved back to Chengdu and only minimal staff remain to feed the ten mournful-looking pandas in their cages. I spend a cold night next to the electric fire and arrange with the driver to take me up the valley the next day.

By dawn, the snow has stopped and the sunshine makes the whole valley look magical. Icicles glisten like chandeliers. Golden Pheasants dash like tiny gleaming dragons in front of the jeep and the white mountains rise steeply on either side like cold ramparts, guarding the secret world of the panda beyond.

The headquarters at Shawan are as deserted as the breeding station of Hetaoping. Giving up the idea of finding any Chinese staff to accompany me to the panda field station, I collect a heavy sleeping-bag from the store, grab some packets of noodles and tinned meat from the tiny shop and we head on another 10 kilometres (6 miles) up the Pitiao valley to halt at the small wooden bridge that crosses the river at the start of the trail up to the research camp.

A cluster of wooden houses remains closed to the cold and smoke drifts through their flattish roofs made from great slabs of slate. There are huge piles of birchwood stacked beside each house, stolen from the nature reserve but without which the Qiang minority farmers could not survive the winter. I order the driver to return in three days, hoist my pack to my shoulders and head off alone up the steeply winding trail.

Despite the deep snow and the cold wind, I am quickly overheated inside my thick jacket and under my heavy pack. But the air is fresh and so clear that I ignore the sweat pouring down inside and enjoy the freedom of being alone in such a wild and beautiful place. The path is steep and twists first this way and then the other. Elliot's Laughingthrushes give clear loud whistles that carry right across the valley. Pheasant tracks criss-cross the path in every direction and the straight-line deep tracks of a fox follow the trail for a hundred metres before veering off into the bushes above. Soon I am high up the side of the valley, and the river and road below look tiny. A young girl swathed in black lumpy clothes takes a large cow along the river bank to find some patch of scrub where it can feed on its tether. The barking of a dog carries up the valley as the last distant sign of human invasion before I am completely lost in the wilderness of the mountain scrub and increasingly dense forest vegetation.

At the top of the valley side the path levels off and I rest for a few minutes under a green pine tree. A flock of pretty little Black-throated Tits flutters busily among the branches of a clump of willows whose crop-curled leaves are held to the twigs only by the ice and snow. Just one more kilometre along a gently rising ridge through thickets of frozen bamboo then, to my left, a small valley drops away and on the far side I can see a wisp of smoke. At least someone is still up at the field camp on this winter's morning. There is no hurry and I take my time, fascinated by the familiar scene under the unfamiliar snow.

Somehow the forest seems sharper and more alive. I can hear further and see further than in the summer and the deep snow leaves clear tracks of the comings and goings of muskdeer, Tufted Deer and tragopans. The path skirts the inside of the ridge and I have to clamber over log bridges to cross two small streams. This trail is on the transition between almost pure conifer forests above, dominated by hemlocks and spruces, and broadleaf forests below, dominated by limes, maples and birches. In winter, these trees are bare and the papery red birch-bark flutters in the wind, gloriously back-lit by the morning sun. Clumps of bamboo form a dense understorey but in many areas they have flowered and stand as dead sticks protecting their tiny green seedlings on the mossy floor, hidden by a protective layer of fresh snow.

The final bend reveals the camp of Wuyipeng, a cluster of canvas tents and two tin shacks that has served as the centre for wild panda research since 1981. Beyond the camp lies a network of trails traversing a study area of some 15 square kilometres (6 square miles). Within the study area, Giant Pandas are lured

OPPOSITE PAGE Wolong Nature Reserve in winter looks cold and inhospitable, but a tale of tracks to be read in the snow reveals that it is still full of animal life. The research camp is almost deserted and great icicles hang from the branches but this is a perfect time to be alone in the forest with the Giant Panda.

RIGHT The Giant Panda does not hibernate but feeds actively on the frozen but evergreen bamboo leaves and stems.

into log traps, immobilized, fitted with radio collars and then followed for the one to three years that the batteries remain good. By now, however, the researchers feel they have learned all they can from the trapping and radio-tracking, and the research has become focused on monitoring bamboo plots and broadening the studies to bears, Red Pandas and pheasants. But today the camp is not full of eager researchers. Like the protection staff of the reserve, they have retreated to winter quarters and I discover only a single guard as watchman living in the camp alone.

I move my gear into one of the empty tents and join the guard for a meal of *dan dan mian* noodles with hot spicy sauce. I spend the afternoon exploring the trails on the west side of the study area. The terrain is terribly steep and I am soon out of breath. I cut a walking-stick from a clump of bamboo to help on the slippery scrambles. There are signs of Giant Panda in several places, but nothing very fresh. More recent tracks, where a herd of Takin has plodded across the study area, are deep and muddy. The bears are all hibernating and as I pass a huge hemlock tree with a great hole halfway up its trunk, I wonder if there is a shaggy sleepy fellow tucked up inside. Broken spruce leaves and fallen twigs on the snow tell me that I have missed a group of Golden Monkeys that somehow live at these altitudes through

the cold winter. It is no wonder the fur of the Golden Monkey is regarded as the best lining for a winter coat in Sichuan. I am angered to find several birch trees in the study area recently cut by villagers for firewood.

I return to camp and plan to look at the eastern end of the study area on the morrow. Dinner is rice and stewed meat with a plate of fried strips of pig rind, spiced with dried red peppers and the strange numbing tingle of Sichuan's favourite spice – *hua jiao*. I report the cut birch trees and hope the guard will chase the matter up. Sleep is hard. Even with all my clothes on and in the thick sleeping-bag I am frozen. A starving rat scurries around the tent looking for my secreted biscuits, preventing any chance of peaceful repose.

The morning is a beauty. The sun shines on the deep snow, crystals of frost twinkle and my breath comes out in steamy clouds. I take the trail towards the prominent white limestone rock that rises above the forest about a kilometre out of camp. A buzzard sits on a bare branch, too cold to bother flying away but alert and watching me suspiciously. I skirt round below the cliff of the rock and see tracks of a serow crossing the scree of a recent landslip. A startled muskdeer raises its long-toothed head to gaze briefly before scampering off into the cover of rhododendrons.

I hear another animal rustling through the underbrush towards me. I stop still and wait. Suddenly a startlingly coloured Scarlet Tragopan comes onto the trail, sees me and waddles off along the path in front of me. He makes heavy progress in the deep snow and I keep up with the bird quite easily. I get a great view of his white-spotted plumage and the blue skin of his face and throat-pouch. I take a couple of photos of the struggling bird and would have followed further but we suddenly cross the fresh tracks of Giant Panda and my interests switch. It is a large animal and I can still smell a faint musty odour on the air. It is heading downhill towards one of the streams of the study area. As there is no trail I use the panda footprints as a guide, placing my own feet exactly in the panda's tracks. Pandas' hindfeet usually land where the forepaws have already trodden, so their tracks look like those of a biped. However, their feet point inwards and I have to waddle in a pigeon-toed manner to take advantage of the panda's trail.

At the bottom of the hill are some large clumps of bamboo. Here the panda has meandered from one to another and broken stems show where it has stopped to feed. In one clump the panda has lain down for a while and there is a big pile of fresh dung, each bolus the size of a goose egg and sweetly smelling of crushed bamboo. However, the dung is already cold, so the panda is still some way ahead. I have a bit of trouble following the trail where the panda has doubled back and crossed his own tracks but soon I am across the stream and climbing up the far slope, and starting to feel a bit like a panda as I waddle from side to side to keep in his footprints. The feeling of being a panda is strengthened when I have to stoop to all fours to negotiate a tunnel through the bamboo. My nose is close to the ground and I can smell my quarry all the way.

At a large hemlock tree the panda has walked around the base and urinated, just like a dog. The snow is stained yellow. Needing to follow his example, I add my own mark to the base of the tree before continuing. The snow is deeper here and the trail shows the groove where the panda has dragged his underbelly along in the snow. The trail is also far from direct, the panda has wandered this way and that, sniffing older trails, marking more trees and occasionally stopping to feed on more bamboo stems. Another pile of dung smells even more strongly and I wonder how quickly it will cool in such a low temperature. Another hollow where the panda has lain down and rolled about. More time lost – I must be catching up with him.

Suddenly the trail turns steeply downhill and becomes a long skid. The panda has tobogganed down the steep slope, only to climb back up the slope again some 50 metres on. Surely he must have been playing, as his descent led nowhere. This is fascinating. I had imagined that in winter, when other animals just hibernate and give up, the panda would be very economical, saving his energy and moving as efficiently as possible from one feeding spot to another. But this panda clearly had energy to burn. He was wandering up and down hills, making detours to visit and mark the biggest trees, playing on the slopes, and feeding a little bit here and a little bit there rather than settling down for a major stuffing.

For myself, the going is very tiring but I know I am catching up and I am excited to get a view of the panda ahead. As I follow along one of our own trapping trails I am surprised to find the panda's tracks divert to enter not just one but two of our log panda traps. When we are trying to catch pandas for collaring it is terribly difficult to lure them into these traps to trip the drop door. The researchers have tried all kinds of baits, with charred goat bones proving the most successful. Even so, pandas are caught only after hundreds of nights trapping effort. This panda

had wandered into two different traps on the same day without any bait at all, just out of interest to see if they were good shelter or to sniff the old smells. I feel we must be doing something wrong in our trapping if we cannot catch such inquisitive creatures.

Finally, I come across what I am looking for. Fresh feeding signs marked by warm, still steaming panda dung. The animal must be quite close now and I move more cautiously forward, still keeping pigeon-toed to the panda tracks. I can hear movement in the bamboo ahead and stop still. The smell of panda is very strong now. I am sure he is only about 20 metres ahead. Suddenly, there is a crashing sound and I hear him running away from me. I hurry after him and he dashes down towards the stream below. It is very steep and I slip and tumble halfway down the hill in pursuit. Now I can hear him still blundering noisily climbing through the bamboo up the far slope. I will never see him now and he will probably continue at that fast pace for at least a kilometre before calming down to his normal routine.

I carry on down to the stream to pick up one of our own study trails to head back home. A short way downstream one of the log bridges has been lost and I have to make a dangerous scramble across the large frozen bluff above the raging and freezing water-pool below. One slip here and I would be in trouble. Panting with exertion, I finally swing across back to the path and head on safer ground for the camp. To the left, under a rock scree, is a salt-lick where a herd of Takin has recently visited for a drink of mineral-rich water. The whole area is ploughed up by their feet like a pig-sty and the fresh dung shows that they were there the previous night. Further along the trail I find tracks of a Golden Cat. A woodpecker hammers in search of grubs in the birch trunks and I start the long climb up out of the valley to Wuyipeng camp and the scent of hot chilli sauce.

I had seen wild pandas on a three occasions before but somehow this long day of putting my feet step by step in the panda's shoes, seeing where he had fed, lain down, urinated, played, feeling like a panda myself as I crept on all fours through dense bamboo thickets, this was the closest experience I ever had with a wild panda. This was the day I started to know what the panda was all about.

AMONG THE ELEPHANTS

Conditions at Sanchahe, in the tropical forests of south-west Yunnan, could hardly be more different from those at Wolong in winter. Few people associate China with tropical forest and elephants, but it was elephants we were after one warm spring morning. Catherine Cheung and myself left the stilted Dai house whilst it was still dark and set off along the narrow trail beside the clear Sanchahe stream, heading deeper into the nature reserve. We were loaded with assorted cameras, water bottles and a snack lunch.

Already, a group of White-crested Laughingthrushes was about and looking for food. They flitted through the low bushes as we approached then burst into a chorus of shrieks and cackles before ending in a trill of maniacal laughter. First light was brightening the sky above but the undergrowth remained dark, and we kept walking into the sticky webs that giant orb spiders had strung across the trail in the night to catch moths and early morning butterflies.

The path wound through a patch of wild gingers and we could smell the aromatic scent of the crushed leaves as we passed. A huge stand of giant bamboos had been twisted this

Asian Elephants (*Elephas maximus*) formerly extended over most of southern China but are now confined to a few tropical forests of southern Yunnan province. Most of the bulls carry heavy tusks.

way and that by elephants and the path had had to be cut again to avoid this great tangle. Piles of dried dung and deep hollow footprints gave evidence of the culprits, but these were several weeks old and we were looking for newer signs. We hurried on along where the trail followed the side of the stream and great trees completely closed a roof above.

We passed the tree root where the pretty Blyth's Kingfisher was wont to perch above the shoal of small fish milling about in the clear pool below. Newly fallen fruit beneath an old strangling fig had attracted a party of junglefowl which scuttled away along the path ahead of us, the cock finally flying up into the trees with a great swishing of wings and a long trail of his glossy tail. In the fig tree above the harsh calls of a Racket-tailed Drongo joined the rising dawn chorus of awakening birds. Patches of mist rose through the trees and a Tokay Gecko added to the morning forest sounds with his wind-up call and deep *toh kay – toh kay*.

We were only 500 metres from our home of the night when we came across fresh elephant signs. Tracks of a mother with a tiny baby and a teenage youngster scarred the hill above, then followed the trail for a while and carried on across the stream. I knew they would visit the salt-wallow ahead, so we did not follow the tracks but kept on the trail and took a short cut to the bend in the river where elephants so often came to drink and to

wallow in a pool of swampy mud. Sure enough, when we got there, the whole area had been trampled over by a group of elephants. There were the tracks of our female with the tiny calf but other animals had been there, too – a larger calf and other females. Piles of fresh dung heaved with movement. I dug in boldly. The dung was already cool but was completely writhing with enormous dung beetles each 5 centimetres (2 inches) long and very strong. A loud buzzing announced a new arrival and within seconds this beetle also was out of sight and burrowng deeply into the fresh food supply.

Puddles of urine smelled strongly of elephant but were already turning brown with oxidization. The elephants had been through in the night and they were some hours ahead of us. We would probably be able to find them, but they might already be some distance away.

Catherine was keen to see them, so we checked our cameras and had a quick leech stop. Later in the summer the whole area would be swarming with these blood-sucking parasites but at this time of year there were only a few small ones attached to our legs. We pulled them off and flicked them away, then set off at a faster pace. Black-headed Bulbuls called above us and a Silver Pheasant scampered off across the trail. We scanned around to pick up the tracks. Different elephants had gone on different routes but I was keen to see the new baby, and I hoped that that family would travel less fast, so as soon as we found the tiny footprints leading away from the wallows down the stream valley we hurried on.

The trail was easy to follow. The ground was soft and the footprints deep. On the steeper hillsides there were deep scars where the animals had slipped and skidded. Twisted palms and a large flattened area showed where the animals had stopped to feed. A little further on we found a clump of bamboo torn and twisted about, another meal and more piles of beetle-heaving dung. The animals had climbed out of the narrow valley and were working gradually along the side of the hill. Their trail dipped down into the shady forest of side streams then climbed again over the intervening ridges where the forests were sparse hardleaf oaks and there was a grassy understorey.

It was a long hot trek. We hurried along at a good pace, both slightly charged by the fresh signs. Here we were not on a regular trail and the path was difficult. The sharp grass and bamboos scratched our legs and arms and the sweat trickled down our backs. The mists had vanished and hot sunshine was beating down into the valley. The elephants must have moved fast. After a couple of kilometres we stopped for a rest. I was sure we were catching up on our quarry but the piles of dung we kept finding were all still cool. The Great Barbets were calling incessantly like an army of hungry cats – *miaow miaow miaow* – and we watched a large black and yellow birdwing butterfly hovering around a climbing vine of the poisonous *Aristolochia tagala*. These butterflies lay their eggs on this plant and the resulting larvae are protected by keeping poisons from the host plant in their own bodies. We had a quick drink and a bar of chocolate but then it was straight off again at the same hectic pace as before.

We found a large flattened area with several dung piles where the animals had obviously enjoyed a prolonged rest. Standing animals often rock from side to side and tread a large area completely flat. From the size of the resting site it was clear that more animals than our own small family group were involved. Maybe they had all met up for a social rendezvous. I noticed that the dung was attracting far more tiny flies than the earlier findings. This was fresher and we must be getting close to the herd. We listened for a couple of minutes but there were no sounds except the calls of birds and buzzing of cicadas. On we hurried, encouraged.

Suddenly, there was a loud snort from a few metres away and we realized we had inadvertently got too close. We could hear an animal moving slowly just ahead but others were moving through the bushes to the side and rear of us. We were right in the middle of the elephant herd. They had been quietly resting and we had not noticed the huge beasts until we were right on top of them.

We stopped silent, listening to try to pinpoint the position of different animals. Although they were very close we still had not seen any. We started moving slowly and quietly back along our own trail but the animal behind us moved again, closing our retreat. I pointed to a tree which had enough branches to climb. It was not very strong but it would get us above the reach of the elephants and it was easy to climb. The larger trees nearby rose sheer for some metres and I doubted we would get up them. Hearts beating fast, we scrambled up into the branches of our tree and caught breath to reorient ourselves.

There was little wind and what there was kept swirling in different directions. The elephants were getting intermittent whiffs of human scent and were moving about nervously, but they did not know exactly where we were so did not know which direction to head. One animal panicked momentarily and rushed noisily for a few metres before halting and snorting. Other animals stood still, rumbling their stomachs loudly. Our view was quite good but the whole hillside was a thicket of dense climber-covered bushes and tall bamboo. From time to time we got glimpses of a brown back, an outstretched trunk and gleam of white tusk or a flicking tail but we still did not get a clear view of a whole animal. I had a small tripod and camera and I waited in vain for a clear shot as a large female fed on a small tree, noisily breaking down one branch after another. A few more paces and she would be in the open but just as she finished she turned the other way and moved back into denser cover again.

A large hornet was attracted to our sweat and buzzed around our heads. We tried to knock it away but did not want to make a noise to alert the elephants of our position. They would probably move away but with young ones present they could get aggressive and they could easily knock down our flimsy tree if they had a mind. The tree was already swaying about as we tried new positions for greater comfort.

Gradually, the feeding animals moved further away and when we were sure there were no stragglers behind us we scrambled down from our perch. Catherine was excited to have seen wild elephants but there did not seem any point in following them further. They were already spooked, were heading further away and we were not likely to get any better views today. Instead, we headed down to the Sanchahe stream. We peeled off our filthy outer clothes and washed them in the clear water, then lay in the shallows cooling off and soothing our scratched limbs whilst our shirts dried on a riverside bush. A large frog swam down the river and I caught it to find it was a beautiful flying frog (*Racophorus*) with elongated hind-toes between which a wide web is extended when the frog leaps from tree to tree to create a buoyant parachute. We let it go and watched it swim off to safety, then we recovered our clothes and wandered back along the streamside trail towards the little Dai house where we were staying.

Over the next few weeks we got to know the individual elephants much better. As part of a WWF survey project we helped to set up five remotely triggered cameras on likely paths in the forest. The cameras were fitted into plastic boxes with glass windows and packed with silica gel to keep out the moisture. The boxes were mounted in tree forks and protected by binding branches and bamboo around them to dissuade elephants from tearing them down and destroying them. Each camera was connected to a small flash and the whole system triggered by a mat buried in the ground. If an elephant or other large animal trod on the mat it would take a photo of itself. The main problem was electric power. The small batteries in the camera and flash would not last on the ready charge for more than a day or so in the damp forest, so a solar-powered panel had to be fixed into the overhead trees to top up a car battery that powered each unit. Fixing the panels in the trees was a major and hazardous labour but one that Mr Dong, a local reserve officer, was easily equal to. A young Chinese researcher, Li Yang, was engaged to check the cameras regularly and change films when necessary. Slowly the photos started to flow in and the animals photographed themselves as planned.

We were surprised to learn that there were far more different animals in the Sanchahe valley than we had imagined. What we took to be the tracks of the same one or two old bulls turned out to be more than ten different bulls, all clearly distinguishable on our photos by their size and tusk shapes. We were also surprised to see what large and splendid tusks some of the males had. In most populations of Asian Elephant there are more females than males and only a percentage of the males have tusks. In the Sanchahe population, we found slightly more males than females and all the males had tusks. We felt this indicated a population that had never experienced heavy hunting pressure. Indeed, the only cases we learned of elephants being killed had been farmers shooting at crop raiders or in fear of their lives. Sadly, this has not remained the case – in 1994 30 men were arrested for operating an ivory poaching ring.

THE PEOPLES OF CHINA

The Han

The peoples of China are varied and numerous. Making up the majority of the massive population are the Han Chinese, who dominate the eastern and lower-lying provinces. The Han are long adapted to high-density living, with rice paddy as their mainstay crop augmented by dry-land farming on hill soils.

Millennia of dense human population have led to almost total transformation of the landscape in these crowded eastern provinces. Almost all natural forests have been cleared, primary forests remaining only on a few mountains protected as sacred sites or of no agricultural use and difficult of access. At the village level, particularly in the south of the country, small wooded hills are left for *feng-shui*, literally meaning 'wind and water' but representing a traditional belief in lucky factors. The *feng-shui* woods are mostly secondary vegetation but constitute significant refuge and both breeding and roosting areas for egrets, herons and other ricefield birds, as well as village birds such as mynas, doves and magpie robins. The other main sources of cover for local wildlife are the dense clumps of bamboo and fruit groves maintained in most Han Chinese villages.

Through long centuries of internal warfare and clan conflict the Hans have developed a gregarious culture, with homes clumped tightly together in villages and towns originally protected by walls. From these, human impacts have radiated into the surrounding countryside, with the wider forest and mountains used as a source of grazing, firewood and hunting.

In many parts of the country coal has become the main fuel but in rural areas firewood is still a scarce and precious commodity. Building timber is in short supply everywhere and has led to heavy deforestation of the countryside. Today, massive-scale reforestation projects are under way to restore productivity to the huge areas of bare hills but the new forests are monocultures of Chinese Pine, Chinese Fir or, in the north of the country, poplars. These are, of course, some cover for birds and other wildlife but nothing like as rich as the original mixed forests that they have replaced.

With little taboo about eating wild animals, the huge Han population has exerted a terrible pressure on wildlife. In the south it is fashionable to eat exotic creatures and people pay high prices to eat owls, monkeys, civets, wild cats and all mannner of other life. In addition, the heavy reliance on traditional medicine adds more burden on local wildlife, with lorises, pangolins, bears, macaques, snakes and geckos, turtles, deer and carnivores all in demand for their supposed medicinal values.

Chinese traditional medicine can be traced back for about 3,000 years and is still the preferred treatment for many types of ailments for most of the people. Its principles are founded on the balancing of the elements: the active male *Yang* must be balanced by the more passive *Yin*, hot must be balanced by cold. Sickness

ABOVE A market in Taipei, Taiwan, specializes in selling medicinal herbs, the raw materials for Chinese traditional medicines. With most plants collected from the wild, rather than being cultivated, the massive trade poses a serious threat to a wide range of species. BELOW Han people at work during the autumn rice harvest in Guangdong province. The Han make up 95 per cent of China's population and are well adapted to high-density living, developing intensive farming techniques and greatly modifying almost the entire landscape.

Girls of the Dai minority picking tea in Xishuangbanna, the tropical region in the far south of Yunnan province. Numbering 840,000, the Dai make up the dominant ethnic group in this border region.

A practitioner of traditional Chinese medicine in Taipei prepares prescriptions, using a colourful mixture of plant parts. Such medicine represents a sizeable proportion of all healthcare.

is an indication that the body is not in balance or harmony and various special foods or medicines are therefore prescribed not so much to treat the specific symptoms of the illness, but to restore the proper balance to the whole system.

The Chinese government has for many years pursued a policy of modernization and integration of medicine, studying and improving the traditional forms of medical practice whilst adopting the best part of western medicine. Most people now accept that many western drugs are more powerful and more direct for treating specific symptoms and accept the value of immunization against common diseases, but find such drugs expensive and distrust their overall effects on the whole body system. Traditional medicines and tonics are often cheaper, considered much safer and provide a more lasting contribution to overall health.

As a result, there is a huge demand for their ingredients. Hundreds of species of roots, bark, bulbs and leaves are gathered daily from forests for personal use or to be sold to drug companies and hospitals. Many plant species are quite endangered by this trade and the human population is now so enormous that the potential demand vastly exceeds the potential supply. In some cases, doctors are experimenting with substitutes for rare ingredients, whilst farmers are also experimenting with cultivation of rare medicinal plants rather than relying on the dwindling supply from the wild. In many parts of China special farms have been established to raise those animal species whose parts are most needed for medicinal supply – bears are reared to get gall extract, civets are reared for their glands and deer are farmed to harvest their young antlers.

As long as China was poor and isolated, this remained a predominantly internal problem but, today, now that the doors are open and China's economy has become stronger than most of its neighbours, this hunger for wildlife has become a huge international issue. The smuggling into China of tiger bones, saiga antelope horn and rhino horn has already given rise to complaints from India and other countries struggling to protect their own wildlife, but trade in turtles, snakes and primates from South-east Asia is now also emerging as a major international problem.

Human impacts on freshwater ecosystems and the marine environment are no less severe than on the terrestrial side. Impounding of lakes, pollution and damming of rivers, overfishing, introduction of exotic species and drainage of wetlands are creating enormous threats to these ecosystems. Many endemic freshwater fish are already lost or heading for extinction, bird species are seriously threatened and some fisheries have collapsed. The Bohai gulf of the Yellow sea, once a famous fishing ground for north-east China, is totally dead as a result of pollution and almost the entire coastline including, in the south, some important coral reefs, is under intense and destructive development.

We may never know what species have already been lost in the eastern provinces but there are certainly many now endangered, including two-thirds of China's total of 285 endangered plant species. Yet, in a way, conditions in the east are more stable than in the rest of China. Loss of natural habitat has taken place over many centuries and the damage has already been done. Species have either survived the pressure of man or not but, today, the remaining forests are coming under protection. Nature reserves are being established, regulations

A young Buddhist monk of the Dai minority works on a bamboo fence at his temple in Menglun, Xishuangbanna. Natural materials are of immense importance to the people of this remote rural region.

Medicinal plant parts for sale in Jinghong, capital of Xishuangbanna. The Dai have an extensive pharmacopoeia based on forest produce, the effectiveness of which is only just being discovered by science.

controlling hunting and trade in wildlife are adopted and the prospects for surviving nature remain optimistic. Some species on the brink of extinction are building back their populations. Yangtze Alligator, Hainan Deer and Crested Ibis have all scraped by and now their populations are once again on the increase. Amazingly, even the South China Tiger somehow survives in the more rugged mountains of this crowded region.

'Bird Loving' week is celebrated in most provinces and schools are starting educational programmes promoting a more understanding approach to wildlife preservation and conservation. Where China's wildlife scene is under the greatest threat is in the less-developed outer provinces of the north-east, south-west and west of the country, where the Han are in a minority and where China's large range of ethnic minorities retain varying levels of autonomy.

Ethnic Minorities

Yunnan is often called the province of the minorities, with more than 30 different ethnic groups across its complex physical geography. Many of these groups are quite small, numbering only a few thousand people. As such, they are freed from the strict birth control regulations that are applied to the Han and larger ethnic groups. Depending on the size of the particular minority, couples can have two, three or unlimited children without penalty. From one point of view this seems a fair approach but, in fact, it is these minority groups that live right on the interface between forest and civilization and it is exactly here that man should seek land-use stability if any balance between artificial and natural landscapes is to be achieved. This is where

wildlife is most threatened and natural habitat is being lost fastest at the present time. For instance, forest cover in Xishuangbanna in south-west Yunnan dropped from 46 per cent to 23 per cent between 1960 and 1980. Similar forest loss is seen in western Sichuan, and even greater loss in Hainan where cover fell from 60 per cent to 7 per cent over the same period.

Two of Yunnan's largest minorities are representative. The Yi, who live from south Sichuan through north and central Yunnan, are hill farmers. Their villages are set on sheltered slopes where their cereal fields form an irregular patchwork of rounded shapes wherever the gradients and soil allow. They are also cattle herders and like to keep some horses. In winter, the cattle roam the secondary scrub and forests in the valleys and on the hillsides, and are fed from the summer straw that is stored in small stacks balanced precariously in trees. In summer, they are taken up onto the highest hills, which have been repeatedly burned to remove forest and scrub and create edaphic grasslands for pasture.

The Yi are unusual in being a matriarchal society, with land being passed down the female line of inheritance. They were also, until as recently as 1947, a tribe of slave owners. Wealthy chiefs and rich families formerly owned most of the land and used the labour of bred slaves and prisoners from other ethnic groups to work the land. Slaves were shackled with wooden harnesses and locked in cages at night. With liberation under communism, slavery was abolished, slaves have become integrated in the tribe and land has been more evenly parcelled out to individual families (after an early stage of state ownership).

The Yi still maintain their traditional colourful costume with broad dark blue cloaks, a rich collection of dances and festivals

and their own distinct language. One of their favourite sports is a type of bull-fighting, where young bulls are set against each other in head-to-head combat and large sums of local money are illegally wagered on the outcome of the bouts.

Further south, the dominant minority are the Dai, who have practised Theravada Buddhism for several centuries in the autonomous prefecture of Xishuangbanna. Like the Yi, they also practised slavery. One of the lesser tribes in the region – the Aini, or Hani – were traditionally a slave class working for the more powerful Dai.

The Dai of Xishuangbanna live in tidy villages, each with its own monastery. Most children still go to school in the monastery where, in addition to the Chinese primary school syllabus, they learn about religious matters and traditional elements of their culture, including care of the environment and respect for nature and living things.

Each village has a *longshan* or 'holy hill', or at least shares a larger one with a neighbouring village. The *longshan*, which may range in size from 4 hectares (10 acres) up to as large as 500 hectares (1,200 acres), is a forested hill where it is forbidden to cut the trees or hunt animals. It is regarded as the home of spirits and it is felt that bad luck will pervade the village if the *longshan* is cut down. However, villagers may rear medicinal plants here and collect mushrooms and other minor products. They may also range their chickens and pigs. Many *longshan* are also used as burial areas. Some are hundreds of years old and act as micro nature reserves against which the heavier impacts on surrounding forests can be gauged. In former times, there were several hundred of these 'holy hills'. Sadly, many were destroyed during the Cultural Revolution and have been replaced by rubber plantations, which have also replaced other valuable rainforests in this part of China.

The Dai have some other interesting traditions. Rather than destroy more and more forest for fuel, they have for centuries planted around their villages fuelwood plots of yellow-flowering *Cassia siamensis* trees, which can be coppiced almost to the ground every few years and soon regenerate. Large clumps of bamboo are also planted around villages to provide shelter, wood, the useful stems that are used to make so many hundreds

of articles and the edible shoots that the Dai like to dry and ferment before using in their cooking. Fences are made around the best bamboo clumps to keep out the village pigs that also enjoy these nutritious shoots.

In the alluvial fans, the Dai make their ricefields, using buffaloes to pull the ploughs and tread the mud. Long lines of colourfully dressed villagers bend to plant out the seedlings and later to cut and thresh the ripe paddy. The Dai live along the rivers and enjoy bathing in these natural waterways. They are not so shy and it is delightful to see parties of topless girls washing their long hair or splashing each other in childish fun, waving to passing boats in the innocence of youth and beauty. The Dai men are great fishermen, using rods, lines, fishtraps and throw-nets. More destructive methods include the use of natural poisons or, if they can obtain them, explosives to stun and kill fish in entire streambeds. Fishing with electricity is a new fad. The men are also active hunters, with two or three muskets stored in almost every household in the district.

Most villages have a 'medicine man' who uses recipes of several centuries to mix traditional medicines gathered from the *longshan* and neighbouring forests. *Dillenia indica* cures stomach upsets, *Dendrobium* orchids are used for coughs, *Tacca* bulbs cure malaria and white cardamom is used in many tonics. The cardamom (*Amomum villosum*) is grown in the forest by clearing the understorey, though leaving the canopy intact, and spreading the rooted stalks of this wild ginger until it forms dense thickets. The seedpods are then harvested and sold in local markets for a high price, together with a whole host of other seeds, roots, animal skeletons, dried lizards and other items considered to have powerful medicinal properties.

On market day, all the young women dress up in their finest clothes to attract maximum attention, and villagers of many other minorities living in the surrounding hills come down into the towns or larger Dai villages to trade and barter. Each ethnic minority has its own traditional costume and can be identified by the colour of the head-dress or type of decoration used. River Dai wear skirts and blouses with broad bonnets whilst the more conservative Black Dai wear black aprons and black headscarves decorated with silver belts and balls. Lahu wear dark blue, Hani

have red pompoms in their head-dress. Indeed, the weekly market is a colourful and interesting scene.

Like most Chinese, the minorities of southern Yunnan follow the lunar calendar. Mid-autumn and spring festivals are celebrated with feasts, copious dangerous distilled liquor or *baijiu* and traditional dances. The spring festival is made even more exciting by the tradition of young girls wandering the streets splashing water on passers-by and, coyly, on the boys they fancy. What used to be a modest dash of water, sprinkled with a flower from a small silver bowl, has become replaced by a drenching with a whole plastic bucketful. This is not the time to wear best clothes and camera splashing is by no means taboo.

Situated on the corner of the famous Golden Triangle, this region of China was formerly a major growing area of opium. The cultivation and trade in opium was completely banned shortly after liberation in 1949 but many of the old villages are still addicted to the 'white powder' which can still be easily smuggled in from over the nearby Myanmar border. In recent years, the control appears to have weakened and trade and addiction to this drug are again on the increase.

The Dai's former slaves, the Aini, are today free citizens but their subordinate culture is still reflected in their habits. Whilst Dai build beautiful spacious villages with large airy houses, those of the Aini are dirty and cramped, with tiny and very

Girls of the Dai minority practise dances in the grounds of a temple during celebrations of the Buddha's birthday. Dai women and girls always dress colourfully, though even more so on festival days, when a veritable fashion show breaks out. Their bright clothes are strongly reminiscent of those seen among their relatives to the south in Thailand.

Tibetan farmers near the Everest north face base camp in the far south of Tibet. Extremely poor, they survive by growing barley on land fed almost solely by Himalayan snowmelt.

poorly designed homes. They have few festivals, wear little adornment and have less agricultural skill. They practise shifting cultivation, know little of irrigation or the use of green fertilizer and have little marketing skill. As a slave class they were never allowed to own land of their own, so simply have no tradition in these areas.

Where the Aini do come into their own, however, is in the forest. If they had no fields to call their own, their traditional refuge was the forest. These are the greatest hunters in Xishuangbanna, shooting the great Gaur and even elephants with primitive muzzle-loading muskets. They also have a superb knowledge of what plants in the forest can be eaten. Camp in the forest with Dai guides and you will have a tough and hungry trip. Take Aini guides and the forest becomes a veritable supermarket of free fruits, vegetables, fish, crabs and other life-sustaining goodies.

As unlike the Dai are from the Aini, so too are the different cultures of Bulang, Yao, Kongka, Jinuo and many more variants who populate these hills and slowly transform the landscape. In the west of China there are many tribes also, but these fall into three main groups: Tibetans, who live on the high plateau areas, Muslim Kazakhs and Uigurs, who occupy the extreme north-west, and Mongolians, who dominate along the northern deserts and steppes.

TIBETANS

Tibetans are hardy people who live in the harshest conditions of the high plateau. In the east the plateau fans out in a series of mountain chains that curve down into Yunnan and Sichuan provinces. Here, Tibetan tribes have adapted to upland farming and are spatially stabilized but in the west of the plateau, in the region known as Jiangtang, the people have no arable land and live in nomadic groups, or *shang*. But, east or west, one thing

remains central to the lifestyle of all Tibetan groups – the yak. Wild yaks still live in remote areas of the Tibetan plateau but the domestic yak is more widespread and is the dominant domestic animal from the west of the plateau, northern Nepal, Bhutan and east into Qinghai, Gansu and west Sichuan. 'Yak' is actually the Tibetan name only for a circumcized male of the species which they refer to as *nor* or *drong*.

Yak wool is woven and then plaited to make the large black tents that form the mobile houses of nomadic families. Yak milk is boiled to make yogurt, cheese and butter so essential for food and barter. Some Tibetans will not kill a yak, but the meat of those animals that die naturally is dried for long keeping and eaten sparingly throughout the year. Dried yak dung provides the main fuel in the treeless wastes of Jiangtang. Yaks are the main draught animals carrying supplies from one campsite to another. Other livestock includes a few horses and some sheep and goats. Sheepskin is the favoured coat to protect the nomads in the harsh winter cold, and the dried dung of sheep and goats is also used for fire although, unlike the less dense yak dung, it needs aeration from a skin bellows to keep it alight.

The staple cereal for most Tibetans is a form of popped barley, eaten dry, mixed with milk or tea to make a broth, or ground into a powder which can again be eaten dry or mixed with tea to make a tacky paste ball. In the eastern parts of Tibet, barley is sown in fields in the short summer growing period but in the west the grain must be bartered for cheese, wool, dried meat or salt. Another item bartered and essential to the Tibetans' lifestyle is tea, a dark red variety unlike that drunk elsewhere in China. The tea is grown in Sichuan and Assam and traded for salt and other products. It is dried into crisp hard balls that can keep for years, and is drunk with butter and slightly salted, stirred with a twiddle stick rotated between the hands. Salt is a precious item so far from the sea, but the Tibetan plateau has many saline lakes and natural salt-pans where Tibetans can go to fill up their yak-wool sacks to trade or for their own use.

Though Tibetans practise Mahayana Buddhism, which differs somewhat from the Theravada Buddhism seen in Xishuang-

Camels carry a Kazakh family's belongings towards lower ground during the autumn migration in the Altai mountains, Xinjiang. Many of the area's one million Kazakhs, a Turkic people, are still nomads, herding sheep and horses.

banna, theirs is in turn a quite unique form of Mahayanism. Like all Buddhists, they believe in reincarnation, but the Tibetans dispose of their dead in what is known as the 'sky burial'. Bodies are cut open or dismembered before being laid out on rocks or a flimsy platform, then crows and vultures descend to do their work of recycling the body tissue into the living universe.

Tibetan society was traditionally feudal, with all land and its human inhabitants falling under the total authority of the supreme Dalai Lama, or spiritual leader, and the Panchen Lama, the second highest leader. Detailed rules and laws dictated who could graze where and who could live where, avoiding overgrazing and guaranteeing a flow of taxes and services to the powerful ruling classes. Following the flight of the Dalai Lama to India in 1959, China tightened its control of Tibet and tried, in the 1960s, to impose a communal farming system on the plateau, but this largely failed and was abandoned. Most Tibetans live in exactly the same way they have lived for hundreds of years, possibly the only way in such harsh conditions.

Although Tibetans have restrictions on killing their own domestic stocks, they are keen hunters, tracking down the Blue Sheep, gazelles and other game with dogs and primitive muskets. Recently, with the advent of roads and long-range rifles, the plateau is being more seriously opened up to hunting and its wildlife is being decimated as never before.

For a few days in the spring Tibetan families gather together for feasting, dancing and songs but for most of the year they drift apart to separate the flocks and use the available pasture as evenly as possible.

MONGOLIANS

To the Tibetans, horses are a luxury. To the Mongolians of the northern steppes, however, horses are a way of life. Man and his horse are almost a single being.

The Mongolians are descended from those great horsemen of Central Asia who throughout history terrorized their more settled neighbours by their speed of movement and their physical bravery. Like the Tibetans, they are nomadic pasturalists, moving their herds from one place to another so as not to overgraze their precious pasture, but while the Tibetans keep within a few days' journey from their bases, the Mongolians take their animals over hundreds of kilometres a year. They herd cattle, horses and sheep and, where suitable, plant cereal crops in the more fertile steppe areas. Whole villages may share a single huge black tent which they move from campsite to campsite.

The early Chinese emperors built the Great Wall to keep these barbaric tribes at bay but when Genghis Khan was able to rally the other Mongol Khans to his banner, and swept his hordes in military conquest both east and west, the Wall proved no obstacle. Within a few years the whole of China was overrun, the Song Dynasty collapsed and the Yuan or Mongol dynasty was established. The new Mongol rulers soon became as Chinese as their conquered subjects and were eventually to be themselves washed away by the Ming dynasty, but in the Mongol homelands life remained unchanged and bands of horsemen patrolled the steppes to protect their herds and families.

At least, all remained the same until very recent times when the Chinese government started a major development of the steppes by establishing croplands in the southern and more fertile regions, fencing the grasslands, sinking boreholes and wells to increase pasture and undertaking mass poisoning of rodents to increase the stocking ratio of sheep. These changes are indeed increasing production, but at a high environmental cost. Rodents are, in fact, essential to the ecological balance of these great grasslands and, faced with fences and poison, many other species of wildlife are being threatened. Even the grass composition is changing.

Overstocking with domestic animals around waterholes has led to serious overgrazing and impacting of soils, so that pasture has deteriorated rather than improved. Species density and mean grass height have decreased, while the proportion of poisonous and inedible plants has increased. Moreover, the fast evaporation around waterholes has led to the drawing up of salt from the ground to cause salinization of surface soils.

MUSLIMS

The Muslim minorities of north-west China grade into the Central Asian tribes of southern Russia and Afghanistan. They are a mixture of races, resembling the Turkish peoples rather than the Chinese and sharing little in culture or language with other Chinese groups. Like their neighbours to the west, they have embraced Islam, rear goats and sheep on the hills and farm the poor mountain soils of this very arid region.

Some Muslims are settled in towns and villages. Young children go off into the hills to tend the herds for days at a time. Other Kazakhs and Uigurs retain a nomadic pastoral lifestyle and live in the famous yurts or cloth tents as they have for thousands of years. In the desert areas, they have domesticated camels and in most areas use ponies and donkeys as the main draught animals. They are a world away from the Li minority farmers of Hainan or the Soshin Indians of the north-east, but such is the scale and diversity of China that it embraces a multitude of nations in one huge country.

THREATS TO BIODIVERSITY IN CHINA

China's fragile environment is facing enormous pressures from its huge population and the needs for development and modernization. In total, these threats can be summarized as habitat destruction, overexploitation and pollution.

The pace of logging has outstripped replanting. Huge areas of forest have been lost in the north-east and in the mountains of southern China, particularly those of the south-west and central regions. New efforts to replant and extend forest cover can again stabilize the area under forest but cannot maintain the biomass of wood or the quality of the original vegetation. Many of these new forests are very young and do not yet constitute dense tree cover. Almost all are monocultures that are ecologically rather sterile. From a wildlife and soil conservation point of view, it would be better to allow natural secondary regrowth rather than to undertake replanting.

Forest fire is a serious problem in many provinces. In the rather seasonal tropical zones of southern China, regular fires gradually degrade and change the natural forests and hinder the restoration of secondary growth. In the coniferous areas, fires can be even more destructive. In China's worst fire, in the extreme north-east of Heilongjiang province in 1987, no less than 3 million hectares (7 million acres) of larch and pine forest were destroyed, with massive loss of wildlife, and the Minister of Forestry was forced to resign. Since that date, fire protection measures have been greatly strengthened. Fires still occur but large-scale damage has been mostly prevented.

Clearing forest land for agriculture, including the practice of slash-and-burn farming in the tropical zone, has been a major cause of habitat loss, as has been the clearing of large areas of good tropical forest in Yunnan and Hainan for the planting of rubber. Similar loss of habitat is seen in the draining of wetland areas for agriculture, the closing of lakes for more controlled management, the development of the coastline and the conversion of steppe and grasslands into fenced farms and ranches. Reduction of water-tables in arid and semi-arid regions, as a result of development of boreholes for irrigation, is causing reduced water levels on many lakes and considerable loss of formerly important wetlands.

Over-utilization affects all resources currently collected from the wild, be it timber, firewood, medicinal plants, fish, or hunted mammals and birds. Of particular concern is the widespread tradition of eating wildlife and using animal species for traditional medicines. The potential demand is simply insatiable and growing whilst the resource base is constantly diminishing. In almost no case can such utilization be seen as sustainable. Collection of rare and exotic wildlife for zoos, exhibition, trade and some so-called but unsuccessful captive breeding programmes are also major threats to some species.

Pollution issues are particularly complex, difficult to solve and increasing in magnitude. China is the world's largest user of coal. Much of this is burned inefficiently giving rise to acid rain and air pollution. Meanwhile, the race for economic growth has spawned the spread of industry and factories far faster than environmental regulations have checked their safety or the cleanliness of the technology adopted. Some terrible pollution is occurring in many of the country's rivers and lakes – all impacting on wildlife and, ultimately, human health and safety as well. Further pollution comes from the development of a major oil industry with resultant spillages. In addition, the South China Sea is one of the world's busiest sea lanes and the dangers of tanker wrecks, leakages and bilging are high.

Perhaps the most unnecessary form of pollution in China is the disastrous pest control policy which results in the spraying of rodenticides over enormous areas of alpine and steppe grasslands, in the mistaken belief that reducing the numbers of rodents such as picas, voles and marmots will increase potential stocking levels of domestic herds. In fact, the poisoning interferes with the entire ecological balance of the grasslands and often reduces their productivity and suitability for domestic pasture. Illegal goldmining also results in local poisoning of water sources with cyanide.

NATURE CONSERVATION IN CHINA

Nature is used and abused on a massive scale in China, as it has been for thousands of years, but the sheer size of the human population now impacts more seriously than ever before. It has been calculated that, on average, every man, woman and child in the country derives over US$300 worth of resources per year in direct and indirect benefit from nature. That means that biodiversity in China is supplying an annual $300 billion worth of services.

At the current rate of loss of forests, wetlands, fisheries, species and services it is clear that this bountiful supply cannot meet demand and will soon be exhausted if direct conservation measures are not quickly put in place. What can be done? And what is the Chinese government doing about it?

During the period of Marxist economy the answer was – very little. Only material and labour inputs were costed into the economic equation. Natural capital such as coal, oil, water, minerals, forests and wildlife were simply seen as free assets or gifts of nature. They were inventorized to see how valuable they were and their exploitation was planned to the best benefit of the state. However, Chinese scientists have, over the past 20 years, slowly eroded such attitudes among economists and, coincident with the opening up of the economy and many other reforms, there is now a growing awareness in China of the importance of and urgency for nature conservation. An early start in developing nature reserves was halted during the Cultural Revolution but has gathered great momentum since 1980. China now has over 700 nature reserves and new ones keep being

RIGHT A Han hunter displays his catch – birds slung across the barrel of his rifle – in the Pearl river delta of Guangdong province. Guangdong is renowned even in China for its exotic wildlife cuisine, the demand for which has resulted in massive persecution of wildlife, endangering a number of species and rendering almost all wild animals extremely timid.

FAR RIGHT A seller of turtles counts her earnings in the infamous Qingping market of Guangzhou (Canton), Guangdong. This market is the main point of sale in the far south for a wide range of wildlife including civets, leopard cats, muntjac deer and various raptors. In recent years, new laws have ended the sale of the most endangered species, though the trade continues.

declared rather faster than central agencies are able to document their details in national databases.

Not one, but three different ministries are engaged in the business of setting up nature reserves and many more reserves have been established by local government at town, prefectural, or county level. For forest reserves, the leading agency is the Ministry of Forestry who have developed some 50 national-class reserves administered from Beijing and over 400 other reserves delegated to provincial forestry bureaux to manage. Many of the most important wetlands and multiple-use reserves and some very important non-forest open terrestrial reserves are established by the national Environmental Protection Agency.

In addition to these *in-situ* strides in nature conservation, China has been very active in establishing breeding facilities for conducting *ex-situ* conservation. Great expense has been incurred to breed Giant Pandas, Golden Monkeys, rare pheasants and deer, bears, tigers, Crested Ibis, alligators and others.

In a few cases, captive breeding has enabled species that have become extinct in the wild in China to be released again into wild habitat. Thus, populations of Père David's Deer have now been established from founder stock returned to China from Europe. The wild deer had disappeared due to habitat loss and overhunting but, during the chaos of the Boxer Rebellion, the French missionary Père Armand David had saved and exported a few animals from a royal hunting park outside Beijing at Nanhaizi. Efforts are currently under way to reintroduce the Saiga Antelope of northern China from Russia and Przewalski's Horse from Russia and Mongolia.

But much more is needed to check the massive over-exploitation of wild lands and species than the mere setting-up of nature reserves and breeding centres. Changes need to be made to the patterns of land-use and resource harvesting. Here, too, China has been active.

The various ministries responsible for the institution of wild resources, forests, fisheries and agriculture have all adopted long lists of laws and regulations to control levels of use and achieve sustainability of the resources in question. For instance the Ministry of Forestry have embarked on monumental reforestation programmes to restore cover and satisfy domestic wood needs from plantations rather than cutting natural forest stands.

Many individual species are protected by law and strict controls are being brought into place to control trade and to penalize those that break the new laws. Some of these penalties are very severe. Several people have already been executed for trading in the skins of that national treasure – the Giant Panda. Moreover, the country is taking its international responsibilities very seriously, too. China has joined the Convention on International Trade in Endangered Species (CITES), is a party to the Ramsar Wetlands Convention and several migratory species conventions, and was only the sixth country in the world to ratify the United Nations' Convention on Biological Diversity.

Slowly, but surely, China takes the right steps and continues to move from principles and regulations on paper to actual implementation. It is easy to find fault and to criticise but, all in all, China is on the right track and deserves a lot of international support and congratulations on its conservation achievements.

The next big step is public awareness. It is one thing that senior leaders now begin to see the great dependence of China on natural resources and the good sense in conserving and managing these resources in a responsible and sustainable way but it is a vastly greater job to get the public to understand the reason for changing their wasteful habits. There is already a steady flow of material on the topic in the news media, some getting across via television, but there is a clear need for putting environmental issues into school curricula and greatly increasing the number of television programmes dealing with such issues.

FOCUS ON
NORTH-EAST CHINA

Formerly known as Manchuria, the north-east unit of China consists of some 1.1 million square kilometres (424,700 square miles) reaching from the Russian border in Heilongjiang and Jilin provinces to the northern parts of Liaoning and including a small part of Inner Mongolia. At the height of the Qing dynasty (1644–1911) the region extended much further north and the great port of Vladivostok was once within its boundaries.

This unit can be broken into four separate physical divisions. The Da (Great) Hinggan mountains form the western flank and the Xiao (Lesser) Hinggan the northern flank, with the Changbai mountains (Changbaishan) along the Korean border marking the south-eastern limits. Between these forested ranges is the plain of the Songhua river, a swampy region of grasslands, lakes and farmland. The Songhua drains east into the Heilongjiang river, which in turn forms the north-eastern border.

Once this was a wild land covered in dense forests, home only to Soshin Indians, and famous for hunting. Today, the north-east is undergoing rapid development and is inhabited by some 40 million people; clearance for agriculture has created extensive plains, and logging has cut a pattern of roads and secondary growth into the hills and mountains. This has become one of the most important grain-producing regions of China, while timber production and petroleum extraction are also major industries.

The climate is strongly continental, with hot moist summers and long cold winters. The extreme northern parts of the Da Hinggan mountains are in the permafrost zone and have a tundra-type vegetation comprising mostly larch forests. Rainfall is about 400 millimetres (15 inches) per year, increasing southwards to about 900 millimetres (35 inches) at the southern limits. The drier northern larch forests are prone to summer fires and in 1987 the worst fires on record in China burned about 3 million hectares (7 million acres) of forest, claiming several human lives as well as destroying wildlife on a large scale.

The Da Hinggan mountains, the highest peaks of which reach over 2,000 metres (6,500 feet), are an ancient feature caused by uplifting of granitic intrusions. The western slopes, composed mostly of larch forest, fall gradually to the Mongolian plateau, while the eastern slopes, clothed in mixed forests of oak and red pine, drop more steeply into the Songhua-Nenjiang plain. Above 1,000 metres (3,280 feet) is a taiga vegetation of spruces (*Picea*), dwarf pine (*P. pumila*) and juniper (*Juniperus dahurica*). At lower elevations, pines (*Pinus sylvestris*) grow on sandier areas and broadleaf trees include such genera as birch (*Betula platyphyla*), poplars (*Populus*), willow (*Salix rorida*) and Mongolian Oak (*Quercus mongolica*).

The Xiao Hinggan mountains are much younger and their landform is rather less dissected. A few peaks exceed 1,000 metres (3,280 feet) but much of the area is rolling forested hills with some swamplands. Forests are generally coniferous, with larch and red pine the dominant species, but towards the south they become more mixed and *Pinus sylvestris* is common on sandy areas.

By far the most diverse forests of the whole unit occur in Changbaishan. These mountains are taller, warmer and moister than the other ranges and so offer much richer habitat for a more complex forest structure and stratification. Coniferous forests include firs (*Abies holophylla*) and spruces, and there are far more genera of broadleaf trees such as lime (*Tilia amurensis*), ash (*Fraxinus mandschurica*) and the dwarf birch (*Betula ermanii*). The alpine zone of the higher peaks contains such species as *Rhododendron*, *Vaccinium*, *Dryas* and *Phyllodace caerulea*.

Over 1,500 species of vascular plants have also been collected in Changbaishan. These include several relict species such as *Pinus koraiensis* and *Phellodendron amurense*. There are several tundra species at higher altitudes including *Linnaea*, *Juniperus sibirica*, *Betula ermanii* and *Rhododendron aureum*. At lower altitudes there are a few elements of the subtropical flora such as *Schizandra clunensis*, whilst the reserve has many rare and endemic plants, such as *Papaver pseudoradicatum*, and the widespread temperate pine *Pinus sylvestris*. In addition, this flora contains over 800 medicinal plants, among them the famous and valuable ginseng (*Panax ginseng*) used to make tea, soups and in various medicines. All told, Changbaishan has the richest and most complete flora in north-east China and this shows a well-marked vertical stratification: mixed broadleaf and conifer forests dominate below 1,000 metres (3,280 feet), conifer forests form a belt between 1,100 and 1,800 metres (3,600–6,000 feet), and birch forests dominate in a narrow zone above this to 2,100 metres (7,000 feet) where, finally, alpine tundra takes over.

In the Songhua-Nenjiang and Sanjiang plains swampy vegetation grows on alluvia that constantly infill the still-subsiding graben landscape. There are many lakes and reedy *Phragmites* swamps but also forested hills and extensive grasslands. In some areas with sluggish drainage the soils are slightly saline and salt-tolerant grass species predominate.

OPPOSITE PAGE Between the forested hills of north-east China lie swampy valleys interspersed with thousands of small lakes – an important area for many kinds of wetland plants and animals. In winter the lakes freeze over, so wildlife must be hardy or go into hibernation to survive, or be able to migrate south.

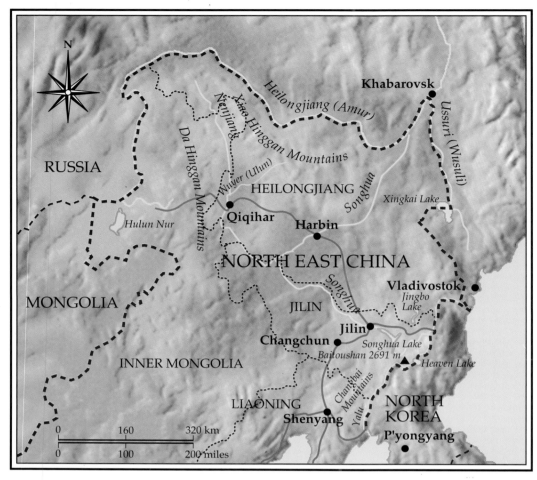

The fauna of the unit is related to that of north-east Russia, and here some tundra mammal species such as Moose, Reindeer and Wolverine reach their southernmost distribution. Other typical northern forest animals such as sable and chipmunks occur and wider-ranging temperate woodland species, including Roedeer, Red Deer, badgers and foxes, are not uncommon. Black Bears roam the forests in the summer months but hibernate in rock shelters through the winter. The tufted-eared Lynx stalks pheasants and hares but the most feared carnivore of all is the Siberian Tiger, which stands a metre (3 feet) tall at the shoulder and can kill the largest ungulates in the forest.

Forest birds consist of typical genera of the northern woodlands – woodpeckers, crows, tits, nuthatches, thrushes, warblers and finches – and a large number of migrant passerines arrive to breed in the area during the summer. There are several pine-seed-feeding species such as hawfinches and the Red Crossbill (*Loxia curvirostra*), an extraordinary bird whose beak tips are elongated and curved past each other instead of meeting tip to tip as in most other birds. This curious shape enables these colourful birds to hook out the slender pine seeds from the cones.

The great swamplands of this unit provide wonderful habitat for waterfowl. There are significant breeding populations of swans, ducks, geese and cranes as well as harriers, rails, waders and other birds that live around the fringes. Zhalong and other important wetlands are frozen solid for several months in the winter and only the hardiest birds remain all year. This is the time when villagers cut the reeds for the paper factories and to make brushes and other household items. Most of the summering and breeding species of the marshes join the Siberian Cranes to swell the wintering populations on Poyang lake and the south coast. But visit Zhalong or Momoge in the summer when the water level is highest and these reserves teem with life. They were set up for the Red-crowned Cranes (*Grus japonensis*) and other cranes that use the area but they also provide sanctuary for other waterbirds.

Many species of ducks, geese and swans use the rivers, swamps and lakes in the summer months, April to June being the main breeding season. Commonest of the ducks is the ubiquitous Mallard (*Anas platyrhynchos*) whose familiar *quack quack* call echoes through the reedy swamps where the downy nests are made on top of a scrape of vegetation. Courting drakes impress the ducks with their bobbing and water-flicking displays and, in almost no time at all, the ducks are paddling around the shallow lakes with their trailing clutch of black and yellow striped fluffy ducklings. By late summer the Mallards form large flocks on safe islands as they moult their flight feathers and are temporarily almost unable to fly. By autumn their new feathers have grown and the young of the year are also flying strongly. Large numbers set off on migration to the lower Yangtze lakes and coastal waters of southern China.

The strikingly patterned black and white Common Shelduck (*Tadorna tadorna*), with its bright red bill and pronounced nob at the base of the forehead, breeds in holes in banks of salty or brackish lakes. Another very common duck is the Common Teal (*Anas crecca*) which has a diagnostic dark green speculum, bordered in front and behind with white. Teal are smaller than the other ducks, have a faster wingbeat and travel at great speed. The drake has a curious cricket-like *kirrip* call and the duck replies with a thin high *guck*. Other common species include the Spot-billed Duck (*A. poecilorhyncha*), named after the yellow spot on the tip of its bill, the Northern Pintail (*A. acuta*) named for its very long pointed tail, and the Northern Shoveler (*A. clypeata*), a large bird equipped with a heavy spatulate bill.

One of the most beautiful of the wetland inhabitants is the Falcated Duck (*A. falcata*). The drake's head is shiny green at the sides and decorated with a long drooping nuchal crest, and his

The forests on the western slopes of the Da Hinggan mountains are composed of almost pure larch (*Larix*). In the drier areas these are susceptible to fire during the summer months, though with improved fire protection measures over the last ten years large-scale damage is more likely to be contained.

tertiary flight feathers extend into long black and white plumes, for which the species is unfortunately rather persecuted. Even more splendid is the Mandarin Duck (*Aix galericulata*). This is the classic Chinese duck celebrated in paintings through the centuries. The drake has a sweeping bold white brow, a golden mane of hackles and beautiful cinnamon display 'sails' that are held erect, concealing the wings. The Mandarin lives on wooded streams in north-east China but many migrate to the south in winter. Unlike the other ducks it breeds in treeholes.

The compact Tufted Duck (*Aythya fuligula*) keeps to reedy ponds and lakes. The drake is black with a white side panel whilst the female is dark brown with some white around the base of her bill. Both sexes have a tuft at the back of the head and a bright yellow eye. Rather than dabbling on the surface like a Mallard, the Tufted Duck makes regular deep dives to feed.

One of the rarer ducks is the Scaly-sided Merganser (*Mergus squamatus*). The male is black and white with a long narrow reddish bill, hooked at the tip. This bill is designed for catching onto fish which the merganser chases under water like a cormorant. The species inhabits fast-flowing mountain streams and breeds in treeholes in north-east China, though it mostly winters further south.

There are several species of geese which live in large flocks and feed on grasses on the drier marshlands and even croplands but rest up in floating rafts on the larger lakes. The Swan Goose (*Anser cygnoides*) is a large long-necked bird with a rather long black bill. The Lesser (*A. erythropus*) and Greater White-fronted Geese (*A. albifrons*) both have grey bills and a pronounced white band on the forehead, while the common Greylag Goose (*A. anser*) is identified by a pink bill and calls with a deep honking as the great V-shaped skeins fly low over the moorlands.

North-east China has three species of swan. The Mute Swan (*Cygnus olor*) can be recognized by its orange bill and characteristic black nob at the base of the forehead. The neck is held in a graceful S-shape. The nest is made on piles of reeds on a few northern lakes and is aggressively defended by the adults until the grey fluffy young are almost full size. The Whooper Swan (*C. cygnus*) also nests in north-east China. The species has a black bill with an extensive yellow basal area. While swimming, the neck is held straighter than in the Mute Swan. The contact call is a loud melancholy bugle-like note. Smallest of the three is the Tundra Swan (*C. columbianus*) which looks similar to the Whooper but has less yellow on the bill. The Tundra Swan nests in Siberia and only passes through north-east China on migration in autumn and spring in large flocks flying in V-formation. Groups chorus with crane-like drawn-out *klah* notes.

The northern lakes are also the home of grebes, storks, ibises, herons, rails and even huge pelicans. The Little Grebe (*Tachybaptus ruficollis*) is a small buoyant bird that lives on reedy lakes and estuaries. Breeding birds utter a shrill trill and make skittish runs over the water. They feed on waterweeds and small animals and dive for prolonged periods under the surface. The nest is made on a floating heap of vegetation and the eggs are covered with debris when the parents are absent. The young hatchlings ride around comically on their mother's back. The much larger Great Crested Grebe (*Podiceps cristatus*) is a more erect and elegant bird with a slender neck and pronounced dark crest. In the breeding season, the adults have a chestnut nape and mane-like ear tufts. Pairs perform elaborate courtship dances, facing each other, rising tall and nodding their heads together, sometimes carrying vegetation in their bills.

Less flamboyant, Common Coot (*Fulica atra*) dabble on the waterweeds, keeping a watch out for the silent harriers that prey on their gawky chicks, and Common Moorhens (*Gallinula chloropus*) slip in and out of the dense stands of reeds.

The Oriental Stork (*Ciconia boyciana*) is a rare endemic bird

and only small numbers still breed in the north-east. With its white body and black wings it looks very like the European White Stork but has a black bill rather than the red bill of the latter. The legs are red and the bare orbital skin is pink. The species is listed as endangered but has been persuaded to breed in some reserves in Heilongjiang by the construction of artificial breeding 'trees', where it can build its characteristic nest of a huge clumsy pile of sticks.

Two species of ibis breed in north-east China. The Black-headed Ibis (*Threskiornis melanocephalus*) is unmistakable, with its hunched stance, white body, black legs, black naked head and sharply decurved bill. It feeds by probing in ponds, marshes and flooded muddy areas. The Eurasian Spoonbill (*Platalea leucorodia*) is taller, with a black spatulate bill. In the breeding season the spatula of the bill is yellowish and there is also a yellowish crest. The spoonbill feeds in shallow ponds, lakes and estuaries with a characteristic sideways sweeping motion of the head.

Grey Herons (*Ardea cinerea*) stalk the edges of the lakes, patiently waiting for incautious fish to move within range of the poised spearlike bill. A sudden stab impales the struggling prey

and after a couple of tosses to get the fish in the right position it is flicked back down the throat. Almost as large, the Great Egret (*Casmerodius albus*) keeps to the shallows and muddy banks, catching crustaceans, small fish and tadpoles. This egret is snowy white with a yellow or blackish bill, and the neck is held in a kinked 'S' shape.

The enormous Dalmatian Pelican (*Pelecanus crispus*) is a rare breeder in the north-east, migrating to winter in the south. It has greyish-white plumage, pale yellow eyes, and the typical orange or yellow gular pouch hanging beneath the huge bill. Pelicans are gregarious and herd fish in groups, dipping their bills in unison, sweeping them to the side and sucking in great pouchfuls of water together with any unlucky fish that are trapped.

During the frigid winter months, the entire Heilongjiang and Songhua rivers freeze solid and only the hardiest birds remain, such as the great Black-billed Capercaillie (*Tetrao parvirostris*), a grouse that feeds on the shoots of the pine trees.

Despite the coldness of winter the summers are very warm and some surprisingly tropical species such as large swallowtail butterflies, reptiles and plump green treefrogs occur in the unit.

OPPOSITE PAGE The rare *Rhododendron chrysanthum* grows around the open banks of Tianchi lake in Changbaishan Nature Reserve. The species has a highly restricted distribution and is listed for protection.

ABOVE The lakes and marshy grasslands of the Taikang area of south-western Heilongjiang are typical of the habitat of the north-east's plains. Though important to wildlife, much land has been lost to oil drilling and agriculture. Today, however, the surviving areas are coming under protection.

RIGHT The extensive grasslands of northern China start in the swampy lowlands of the north-east, then extend through the steppes of Inner Mongolia for thousands of kilometres across flat landscapes. This is the land that produced the warrior horsemen of ancient China.

RIGHT A White-naped Crane (*Grus vipio*) flies over the swamp grasslands of the Zhalong Crane Reserve in Heilongjiang – a symbol of the wildness, freedom and space of this north-east frontier.

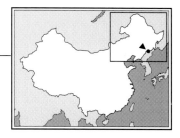

Changbaishan

Changbaishan Nature Reserve is located in the south-east corner of Jilin province on the border with North Korea. The area was made a reserve as early as 1961 and benefits from extensive inventory work and studies by scientists of the Chinese Academy of Sciences. It is also a Man and the Biosphere Reserve under the UNESCO MAB Programme.

The reserve is in a volcanic zone and the main mountains are composed of volcanic cones and lava flows. The highest peak, Baitou, which reaches 2,691 metres (8,282 feet), is in fact the highest mountain in north-east China and is the source of the Songhua, Tumen and Yalu rivers. Nearby is China's largest crater lake, Tianchi, lying at an elevation of 2,100 metres (6,800 feet) and over 200 metres (650 feet) deep at the centre.

Changbaishan is very rich in flora and fauna. Over 50 mammals have been recorded here, including the rare Siberian Tiger, Leopard, Sika Deer, Goral, Red Deer, Lynx, otters and sable. Little chipmunks are common on the ground, inquisitively coming close to visitors then flicking their tails and dashing off with a derisory *chonk* cry. More than 200 species of birds are known from the reserve including such rare species as the beautiful Mandarin Duck, Oriental Stork and Scaly-sided Merganser.

Changbaishan extends over the border into North Korea and would form a good example of an international transfrontier reserve if both countries could co-operate on its management and protection.

Close to Tianchi (Heaven lake) the upper slopes of Changbaishan include numerous dramatic cliffs, the outer walls of a volcano whose crater holds the lake's waters.

ABOVE The conifer zone of Changbaishan, which lies between 1,100 and 1,800 metres (3,600–5,900 feet), is dominated by pines. The vegetation is lush and the understorey is clothed in tall ferns; these are the richest conifer forests in north-east China.

RIGHT The pretty Scarce Swallowtail (*Papilio xuthus*) looks large and tropical but is one of the hardier temperate butterflies, extending to the very north of the country.

BELOW LEFT The Japanese Robin (*Luscinia akahige*) breeds in Siberia but passes through north-east China on migration to the south for the winter. It has a red breast like its European counterpart but is only distantly related.

BELOW RIGHT The decorative Mandarin Duck (*Aix galericulata*) breeds in tree-holes along forested streams of north-east China, and winters in the south as far as northern Vietnam, but it is nowhere common. It is a favourite of captive collections all over the world

ABOVE At an altitude of over 2,000 metres (6,500 feet), right on the border with North Korea, summer around Tianchi is extremely short. Even in July, ice still floats on the surface of the lake, a reminder that for nearly nine months of the year it is frozen over.

OPPOSITE PAGE Water flowing out of Tianchi soon cascades down in a spectacular waterfall. In winter, the extremely low temperatures cause the fall to freeze in motion in a magnificent sculpture of ice stalactites.

Pretty indigo gentians (*Gentiana*) flourish in the moist vegetation of Changbaishan's alpine zone.

Dwarf wild *Aquilegia* grace the forest floor. The horn-shaped lobes hold copious moisture for hardy northern moths.

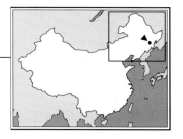

Jingbo Lake

Jingbo is a large freshwater lake formed by the eruption of several volcanoes which dammed the Mudanjiang river to leave 40 kilometres (25 miles) of deep water surrounded by primary and secondary forests. The indented shape of the lake leaves many secluded sandy bays, and a boat-trip through this lovely waterway reveals new and exciting views around each bend. Rugged cliffs and rolling forested hills that throw shadows and reflections provide some memorable landscapes.

The lake is used as a staging post for migrating flocks of waterbirds and has a resident population of Little Grebe, but is otherwise surprisingly lacking in birdlife. It is the scenic splendours that attract thousands of visitors each year.

There are extensive forest areas north of Jingbo, which are very beautiful with a mixture of red pines and larch, and many broadleaf trees such as oaks and maples. These forests are alive with the chirping of birds and squeals of the little striped Siberian Chipmunks that scout the tourist trails cheekily looking for leftovers. Hawkers sell rare fungi and ferns that have highly recommended medicinal values.

The extinct volcano of Huoshankou offers some amazing scenery and is a great attraction for domestic visitors. But there is also important wildlife in the area. Tracks of Siberian Tiger were recorded in the reserve as recently as 1987 and there are Leopard, Lynx, Muskdeer, Red Deer, Sika Deer and Goral to watch out for in the remoter parts.

Jingbo, a beautiful lake in the south of Heilongjiang, was formed when a valley was dammed by lava flowing from a nearby volcano. Its surrounding forests, including those on the slopes of the now extinct volcano, are similar to those of the Changbaishan region.

The mixed forests around Jingbo are very rich. Red Pine (*Pinus koraiensis*) grows among a host of broadleaf trees including maples, oaks, limes, elm and hazel. Tits and flycatchers feed busily and chipmunks squeal and chase one another among the trees.

The red flowers of *Lilium* aff. *concolor* add a bright splash to Jingbo reserve's symphony of summer greens.

The large yellow trumpet of a graceful day lily (*Hemerocallis*). Local people cook the buds as a vegetable.

These glossy bracket fungi (*Ganoderma*) grow on pine trunks and look totally inedible but are used to make medicines.

Despite winter temperatures of many degrees below freezing, Heilongjiang has long warm summers and even such heat-loving creatures as these plump green treefrogs (*Hyla*) are able to survive.

Chipmunks (*Tamias sibiricus*) are common in the forests around Jingbo. These alert, active animals spend much time on the ground feeding on seeds, nuts, fruit and fungi.

The Silver-washed Fritillary (*Argynnis paphia*) is a common insect of the northern forest glades through the summer months. It has both a dark greenish and paler rufous form.

The Bath White butterfly (*Pontia daplidice*) has a wide distribution all the way to southern Europe. In north-east China it frequents the flowers and herbs of open grassy areas.

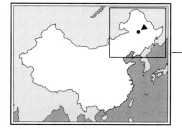

Zhalong Crane Reserve

Zhalong was established in 1979 to protect important wetlands and the breeding grounds of the magnificent Red-crowned Crane. The reserve is very large with an area of over 2,000 square kilometres (770 square miles) composed of lakes and marshes in the lower drainage of the Wuyer (Ulun) river. There are extensive reedbeds of *Phragmites communis* and some of the seasonal ponds are slightly brackish. The wetlands support a wide range of aquatic plants and the reserve, together with the surrounding grasslands and cultivated areas, provides a haven for all kinds of waterfowl.

Cranes and swans nest on great piles of vegetation heaped up in the reedbeds. Largest and most elegant of the inhabitants are the Red-crowned Cranes for which the reserve was created. About 200 of these graceful birds summer at Zhalong. In addition to the wild breeders the reserve undertakes a captive breeding programme with the assistance of the International Crane Foundation. The cranes usually have two yellowish chicks, but swans may have 8 to 10 dowdy chicks per brood. Other important breeders include spoonbills, cormorants, bitterns and many species of ducks and geese. In total 250 species of birds have been recorded in the reserve.

Mammals in Zhalong include Roedeer, foxes, badgers, martens and the occasional Mongolian Gazelle on the neighbouring grasslands. There are also a few turtles, amphibia and some 42 species of fish.

Zhalong is a flat swampy area of *Phragmites communis* reedbeds and grasslands interspersed by shallow, slightly saline lakes. It is the most famous wetland in north-east China and the principle breeding area for cranes and many other waterbirds.

ABOVE White-naped Cranes (*Grus vipio*) stalk the lake banks in search of juicy tubers on which they like to feed. They will also take insects and small vertebrates.

LEFT AND BELOW The Red-crowned Crane (*G. japonensis*) is China's largest and most splendid crane. Adults stand about 1.5 metres (5 feet) tall. Zhalong was especially set up to protect this endangered species and about 200 of these magnificent birds live in the reserve. There are usually two chicks per brood.

ABOVE An *Orthetrum* dragonfly of the family Libellulidae takes a rest on a small tree. It makes sudden darting flights to catch smaller insects in mid-air but must itself watch for the eager bills of the many waterbirds.

The Grey Heron (*Ardea cinerea*) is one of Zhalong's inhabitants and is found along streams and lake shores throughout northern China, patiently stalking and catching small fish and frogs in the shallows.

A stripy frog (*Rana nigromaculata*) thrives in the vast marshlands. These creatures hibernate through the winter but croak the nights away during the summer months when the water rises.

FOCUS ON NORTH CHINA

Landform varies in this region of hills, plains and small mountain ranges. The fairly level Loess plateau lies at about 1,000 metres (3,000 feet) in the west, the lowland plains of the Huanghe (Yellow river), Jing and Luo rivers spread in the north, while along the south run the mountain ranges of Qinling, Daba, Micang and Dabieshan. At the eastern end the rocky Shandong peninsula extends as rolling hills into the gulf of Bohai.

The north China plain is composed of the huge amounts of silt and sediment brought down by the Yellow, Jing and the Luo rivers. The Yellow river, in particular, collects heavy silt from the Loess plateau and this is deposited in such large quantities that the river has changed its course many times through history, often causing terrible floods with massive loss of human life. The pattern of floods has brought this area great hardship but enormous efforts have been made to build channels and embankments and, as a result, the river has kept more or less to the same course since 1855, with the last minor shift in 1976. The north is now one of the most important agricultural areas of the

country. There is almost no natural habitat or original wildlife in the transformed landscape.

The infamous Loess plateau, stretching over an area of more than 500,000 square kilometres (19,300 square miles), takes its name from the deep layers of wind-blown yellow dust that have accumulated over millions of years from the deserts of northern China. The dust is lightly compacted into a porous, silty soil, rich in nitrogen, potassium and phosphorus which give it great fertility. However, because of its loose structure, the loess is very prone to erosion and the landscape is crenellated by thousands of small winding gulleys. Farmers have tunnelled their homes into the sides of these gulleys but when heavy rain falls, these gush

OPPOSITE PAGE Late autumn in the birch and oak forest of Wulingshan Nature Reserve, a mountainous area just inside Hebei province to the north-east of Beijing. Such forest is typical of the region's natural vegetation, but little survives outside protected areas due to pressure from the human population.

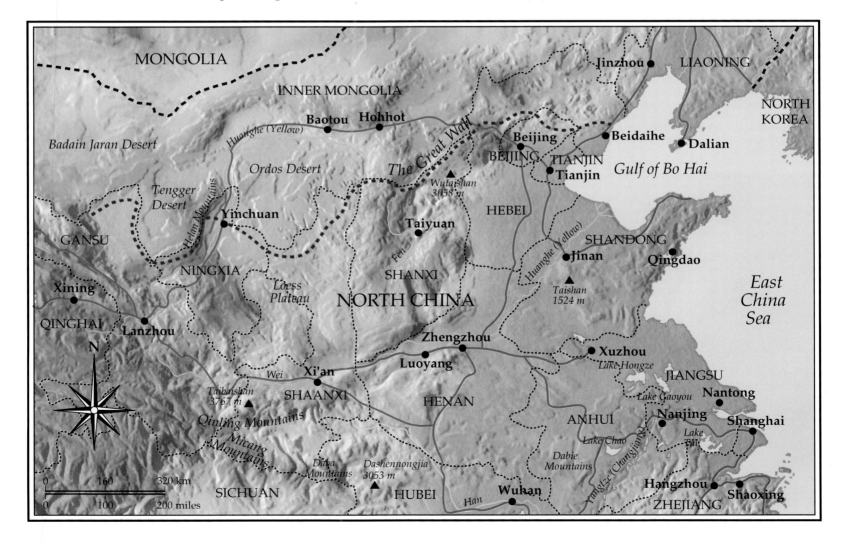

The Oriental Stork (*Ciconia boyciana*) is a rare breeder in China but at Beidaihe, on the Bohai gulf, large numbers may be sighted on migratory passage to the south.

ABOVE The Common Otter (*Lutra lutra*) has a huge distribution but is everywhere persecuted, both because it depletes fish stocks and for its valuable fur. OPPOSITE PAGE Two captive herds of the handsome Père David's Deer (*Elaphurus davidianus*) have been reintroduced into China.

with yellow, silt-laden water which is all channelled down to the Yellow river. On a windy day, the dust is raised into the sky to create spectacular yellow clouds that carry the dust further east. Clearance of the original vegetation has added greatly to these problems. Much of the effort to control the flooding of the Yellow river consists of reforestation and planting grasses over extensive areas of the plateau.

The Yellow river is China's second largest river after the Yangtze. It rises in the Bayan Har mountains in Qinghai province and flows some 5,500 kilometres (3,400 miles) through nine different provinces before discharging into the gulf of Bohai. In its upper reaches the river is clear and opens out into a series of lakes and grassy swamps. Its course runs north-east through Gansu province, passing the capital, Lanzhou. On either side the landscape becomes a desert, though both banks are bordered by a green swathe of agriculture and plantations. The river looks deceptively placid but soon the channel becomes more lively as it passes into the Loess plateau through a series of rapids. The most famous and spectacular of these are the Hukou falls where the river, already stained yellow from the Loess silt, tumbles 17 metres (55 feet) and is compressed from a width of 250 metres (820 feet) through a channel of only 50 metres (160 feet) to create a churning, swirling cauldron.

As the river passes through its middle sections it collects the bulk of the sediment which gives it both its colour and its name, then for the last 900 kilometres (560 miles) runs smoothly across the north China plain, depositing much of its silt and raising its bed a few metres above the surrounding countryside. The river itself is fanned out and shallow, making navigation by boat difficult; cargo ships cannot pass and sailing vessels must keep to the ever-changing deeper channels. Finally, the river discharges into the sea, creating a huge apron of yellow fresh water that can be seen from space to extend far out into the Bohai gulf.

Separating the Bohai from the larger Yellow sea are the low hills and mountains of the Shandong peninsula. Most of these hills are below 500 metres (1,600 feet) but a few peaks, such as Taishan and Laoshan, rise to over 1,000 metres (3,280 feet). The peninsula has beautiful scenery and a convoluted, rugged coastline. Most of the land has long been cleared and modified by man but there are some small natural forests of oak and Japanese Red Pines among the pretty mosaic of orchards and farmlands.

The Qinling to Dabie mountain chain forms the southern edge of the north China unit and the watershed between the Yellow and Yangtze rivers. These mountains are also a major barrier between the subtropical forests to the south and the warm-temperate deciduous forests to the north. Conditions on the two sides of the mountains are quite different: on the south the forests are sheltered, warm and moist, inhabited by tree-frogs, porcupines and bamboo rats; to the north the slopes face harsh winters and cold fronts, and they are dominated by conifer forest above a narrow broadleaf temperate zone. Here are found bustards, jumping-mice (Zapodidae) and sand-grouse (Pteroclididae).

The north has been the economic, cultural and political centre of China for hundreds of years and is the most highly populated region, supporting over 250 million people. As a result, little of its original vegetation of mixed temperate broadleaf deciduous forest now survives. Most of the lowland forest has been cleared over the centuries to make way for orchards and the great fields of wheat and cotton. Natural forests remain only around Buddhist temples and in some nature reserves, such as the Qinling mountains. This is particularly sad from a conservation viewpoint, as these forests are far richer in both genera and species than any comparable forest found in Europe or the eastern United States of America.

The dominant tree genus is that of the deciduous oaks (*Quercus*) with mixed elements of elms (*Ulmus*), wild walnuts (*Juglans*), maples (*Acer*), nettle-trees (*Celtis*) and ashes (*Fraxinus*). A few conifers, such as larches (*Larix*) and Red Pine (*Pinus koraiensis*), are also found and birch (*Betula*) is an important forest tree at higher altitudes.

The Qinling mountains show the most clearly defined altitudinal zonation. Lowland forests are mixed evergreen and deciduous broadleaf with Sawtooth Oak (*Quercus acutissima*) and Chinese Cork Oak (*Q. variabilis*) as the main deciduous species. Chinese Pine (*Pinus massoniana*), Blue Oak (*Cyclobalanopsis glauca*) and Bitter Chinquapins (*Castanopsis sclerophylla*) are the main evergreen components. At higher altitudes, between 1,200 and 2,600 metres (3,900–8,500 feet), the forests are mixed broadleaf with conifers. Armand Pine (*Pinus armandii*) and East Liaoning Oak (*Quercus liaotungensis*) dominate the canopy, with increasing incidence of Chinapaper Birch (*Betula albosinensis*) and Hornbeam (*Carpinus turczaninowii*) as one rises through this zone. Between 2,600 and 3,000 metres (8,500–9,800 feet) conifers

dominate, with Farges's Fir (*Abies fargesii*) and the endemic Shensi Fir (*A. shensiensis*) as the main species. Above is a zone of Chinese Larch (*Larix chinensis*) with an understorey of Autumn Purple Rhododendron (*Rhododendron fastigiatum*).

On the open steppes, grassland vegetation extends as far as the eye can see. Main species components are *Stipa grandis* with *Aneurolepidum chinense* or *Stipa krylovii* with *Cleistogense squarrosa*. Once, great herds of Mongolian Gazelle (*Procapra gutturosa*) roamed these grasslands but those times have gone. What has not been converted to wheat production is now grazed by cattle, sheep and horses.

The fauna of the unit is characteristic of the temperate forest and does not differ so much from western Europe. The sparse forests harbour deer, wild pig, red fox, badgers, hedgehogs, pheasants and red squirrels. If you are lucky you may hear the deep *boop* call of the huge, amber-eyed Eurasian Eagle-Owl (*Bubo bubo*). By day the forests are alive to the hammering of the Great Spotted Woodpecker (*Picoides major*). Eurasian Nuthatches (*Sitta europaea*) clamber down trunks and branches searching for insects or nuts and the delicate tree-creepers (*Certhia*) work upwards. The Great Tit (*Parus major*) gives its harsh *cher cher cher* call and thrushes and blackbirds flick through the leaf litter on the forest floor. Jays (*Garrulus glandarius*) and the Black-billed Magpie (*Pica pica*) are seen frequently but the Red-billed Magpie (*Cissa erythrorhyncha*) is one rather splendid addition that is not found in Europe. It has a bright red bill and a long white-tipped tail to go with white underparts and a bright blue back.

Another crow – the Azure-winged Magpie (*Cyanopica cyana*) – is rather interesting. This pretty bird, which lives in flocks, has a smart black cap, white underparts and grey-bluish upperparts. In autumn, when the persimmons hang ripening in Chinese gardens and orchards, these magpies make concerted attacks on the sweet fruit and the air is filled with their soft cries. What is most strange about this bird is its global distribution. It is common in the north China unit, then totally absent from all intervening lands until it is again found as a common bird in southern Portugal and Spain. There is no documentary evidence to suggest that the bird was ever introduced to the Iberian Peninsula, but such a disjunct distribution is very unusual in animals and it does seem most likely that some early Portuguese sailor must have brought back a cage of these pretty crows, which escaped to colonize a new but ideal habitat.

Main raptors of the forests and woodlands are the Common Buzzard (*Buteo buteo*), Northern Goshawk (*Accipiter gentilis*) and Eurasian Sparrowhawk (*A. nisus*). The buzzard uses the forest for nesting, and sometimes raids other birds' nests, but mostly hunts over open country. The other two, however, are primarily forest hunters. Their wings are relatively short and rounded, allowing them to make the sudden changes of direction and accelerated bursts of speed necessary to outpace their prey around the obstacle course of trunks and branches. The sparrowhawk catches small birds whilst the larger goshawk can take squirrels, hares and pigeons. Both species nest in large twig piles in conifer trees. The Northern Hobby (*Falco subbuteo*) is a small falcon that flies swiftly over the treetops and clearings, catching dragonflies and other insects in mid-air. Occasionally it even catches swifts or other small birds. The hobby often makes use of an old crow or magpie nest.

Wildlife on the steppe is quite different. Several species of larks nest on the grasslands and sing magically on fluttering wings high in the sky. The Great Bustard (*Otis tarda*), the world's heaviest flying bird at up to 15 kilograms (33 pounds), walks majestically over his territory and nests on mounds of dry grass. Wagtails, wheatears, accentors and the Pink-tailed Bunting (*Urocynchramus pylzowi*) are common in places. Daurian Partridge (*Perdix dauuricae*) live in small flocks on the higher ground, Carrion Crows feed on the open fields and Northern Lapwings (*Vanellus vanellus*) wheel in chaotic flocks, shrieking at imagined dangers. Voles and marmots go about their daily chores in underground burrows.

The great Steppe Eagle (*Aquila nipalensis*) makes its nest in an old marmot burrow and terrorizes its neighbourhood by sailing slowly over the open land, staring fixedly at the ground below. As winter approaches the eagles gather into groups and migrate to the south and south-west. Other raptors include the Peregrine Falcon (*Falco peregrinus*) and Common Kestrel (*F. tinnunculus*). The first chases its prey in the air and catches them by sheer superior speed. The kestrel, however, takes mostly small mammals and can hover stationary on fluttering wings, watching and listening for the tell-tale movements that signal dinner and trigger a stooping descent with outstretched talons.

The region is the northern limit for monkey distribution in China. The unit is generally low in endemic animals but does boast the ornate and now rather rare Brown Eared Pheasant (*Crossoptilon mantchuricum*) and the Père David's Deer (*Elaphurus davidianus*). The latter species formerly roamed wild over much of the unit but was hunted almost to extinction. Through captive breeding, a herd was established at Dafeng Reserve on the east coast and a second at Nanhaizi, outside Beijing. Both herds have bred up to several hundred animals and a third wild population is now being established in Hubei.

The fate of the Brown Eared Pheasant was almost as precarious. The species has magnificent plumed tail feathers, which in Imperial times were traditionally awarded to victorious generals as a symbol of royal thanks for valour. Hunting for these plumes almost drove the pheasant to extinction but small populations of survivors were found in Hebei and Shanxi provinces, and two nature reserves were set up specifically for the species. Luckily conservation arrived just in time and the populations are once again on the increase.

The unit's coastline is an important route for migrating birds. Several reserves have been set up in Shandong to provide feeding habitat for the large migrating flocks, and a monitoring post and reserve are run at Beidaihe, where spectacular sightings of rarely seen species can often be made. In 1986 almost 3,000 of the rare Oriental Storks (*Ciconia boyciana*) were seen here, which is a large proportion of the world population.

To the south of the Qinling mountains the forests continue uninterrupted to the subtropical zone where at Yaojiaogou, in Yang county, remains the last spot in the world where the curious Crested Ibis (*Nipponia nippon*) still survives in the wild. This pinkish-white ibis has a bare red head, decurved red bill and a long floppy crest of pink plumes. Formerly the species ranged as far as north-east China and Japan but the population has been reduced to a few pairs, all breeding in a few ancient oak trees close to farmland. The birds feed in open wet places, including cultivated fields, and need intense protection from hunting. In addition to trying to protect the species at this site, a joint project with Beijing zoo has taken a few birds into captivity and has had some success in captive breeding of this precious species there.

Leaves take on autumn hues in Wulingshan. The reserve protects a rare remnant of mixed broadleaf and conifer forests native to the region. In contrast, the distant hills are bare or patched with planted pine monocultures.

Qinling Mountains

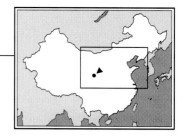

The Qinling mountains are very important for conservation. They shelter the Takin, which has a rich reddish or golden form here, as well as large populations of Golden Monkeys. This is now the eastern limit for the distribution of Giant Pandas, which have been studied for many years in the excellent nature reserve at Foping. But probably the most important reserve in the Qinling area is the sacred mountain of Taibaishan, which rises to 3,767 metres (12,359 feet) and is situated only 120 kilometres (75 miles) south-west of the famous Shaanxi capital of Xi'an – a former capital of China and the home of the amazing Terracotta Army.

Taibaishan is a large reserve representing both the north and south faces of the Qinling range. A narrow trail winds up the mountain from the north to the very summit, where a small freshwater lake remains almost lifeless and coated in ice even in midsummer. The path descends again on the other side to the south. Along the trail are small monasteries, 28 in all, where Taoist and Confucian monks go into retreat to contemplate in the quiet mountains and where, at certain times of the year, pilgrims from a wide radius come to pray. Centuries of such religious guardianship have protected the forests, its trees and animals from hunting and destruction, making Taibaishan in fact one of China's most ancient nature reserves.

Today, the increasing influx of tourists and the noisy and intrusive developments that always accompany tourism – transport, radios, food vendors and litter – are disturbing the tranquillity. This is a problem that must be solved if the mountain is to retain both its value as a site of religious meditation and its importance for biodiversity conservation. A new conservation project is being set up to work closely with the religious authorities on a management prescription plan for the reserve and suitable regulations for its use.

The Qinling mountains are a mixing ground of northern and southern Chinese elements, showing a wide range of vegetation types, from the subtropical evergreen forests on the south to cold conifer forests with dwarf bamboos on the higher ridges. Giant Pandas, Golden Monkeys and Takin are all sheltered here.

A rich variety of wildlife and plants is sheltered in the Qinling range. RIGHT The Velvet-fronted Nuthatch (*Sitta frontalis*) occurs throughout the tropical and subtropical zones of China. It searches for insects along the trunks of trees. BELOW LEFT In the wetter parts of Taibaishan Nature Reserve the forest floor is carpeted in a dense layer of shade-tolerant climbers. BELOW RIGHT The Carrion Crow (*Corvus corone*) has a range extending from Europe to northern China. BOTTOM LEFT In the autumn, Sambar stags (*Cervus unicolor*) challenge each other for the right to females but mostly the deer remain solitary and only rarely form herds. BOTTOM RIGHT A scaly lizard (*Takydromus*, family Lacertidae) clambers among the summer vines. It will hibernate through the winter.

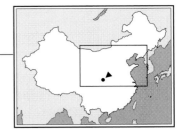

Shennongjia

The Shennongjia prefecture, in the west of Hubei province and to the south of the Qinling mountains, has some of the richest mixed temperate broadleaf forests in the world. In autumn, the trees turn to gold, enhanced by the yellow of larches and the deep red of beeches.

The nature reserve was formerly a timber production farm and the most accessible forests have been cut. But some primary forests remain at highest altitude or in steep valleys and around the tall limestone pinnacles, where mists constantly swirl and shroud the landscape. These pinnacles halt the flow of moisture to the west but the winds soon lift the clouds over the ridges of Shennongjia to send them on to the higher peaks of Qinling and Min. The conifer zones of the upper levels of the reserve are carpeted by dense bamboo thickets and in some areas, where fires have destroyed the sheltering forest canopy, there are open areas of pure bamboo.

Outside the reserve, broadleaf forests stretch away in extensive rolling hills. Oaks and beeches predominate but chestnuts, maples, larches and even some subtropical species such as *Lindera* occur. These forests are the easternmost limit of Golden Monkeys, which live in large groups, moving through the subalpine conifer zone for most of the year but coming down into the deciduous forests in winter.

Giant Pandas are now extinct in Shennongjia but other important wildlife includes Black Bear, Leopard, Serow and such temperate species as foxes and badgers. The reserve is proud of the strange abundance of albino forms that have been recorded here, including pheasants, Golden Monkeys and even a pink-eyed, white Black Bear.

Most extraordinary, however, are the stories and signs of the Hubei wild man – a type of yeti that supposedly lives in these mountains. Hairs, dung and footprints of this 'abominable snowman' can be seen in the reserve museum and the Chinese government has set up a cash reward for anyone who can provide conclusive proof of the existence of this highly doubtful creature.

OPPOSITE PAGE The Shennongjia Nature
Reserve protects the highest crags and peaks of
a largely forested prefecture. A local yeti, or
wild man, is said to inhabit these hills but the
hairs shown in proof are those of the Golden
Monkey. Misty pinnacles provide mysterious
enough scenery for anything to be believed.

RIGHT The Golden Monkey (*Rhinopithecus
roxellanae*) has brilliant reddish-gold fur and a
remarkable blue face with an upturned snub
nose and lappets at the side of the mouth. The
species lives in conifer forests up to 3,000
metres (10,000 feet), coming down to 1,800
metres (6,000 feet) in winter.

BELOW LEFT Berries in autumn (*Cornus*) shine
scarlet to attract seed-dispersing birds.

BELOW RIGHT Asiatic Black Bears (*Ursus
thibetanus*) are widespread in China. They
hibernate in winter but scour the forests for
fruit and grubs in summer. Bear gall is used as
medicine by the Chinese, but is now mostly
taken from animals reared on special bear
farms. Since 1981 the Black Bear has been
totally protected.

Fragrant Hills

In a region of China where there is almost no natural vegetation left, the patches that do persist are almost all scenic sites well trodden by the feet of visitors from one of the most densely populated regions on earth. An especially well-frequented area is Xiangshan or Fragrant Hills, a natural park and botanic garden on the outskirts of Beijing. In October, the temperate woodlands become a mass of autumn colours with scarlet aspens, yellow poplars and a few bright red maples mixing with the browns and orange of oaks, larches and other deciduous trees.

Thousands of visitors stream to the hills to enjoy the *hong ye,* or red leaves, and hawkers sell maple leaves sealed in plastic sheets as tourist souvenirs. The few squirrels, birds and an occasional fox hide in the densest cover until the winter frosts snap off the last leaves and the cold sharp weather drives away the tourists for another year. Colourful fungi are trampled by clumsy shoes and hardly a square metre of the reserve escapes discovery by the noisy bipeds. As winter closes in flocks of

Azure-winged Magpies attack the last scarlet persimmons hanging in bare trees and the larger Black-billed Magpies give cackling cries beside their complex, domed twiggy nests that stand out conspicuously in the bare poplar crowns.

With the arrival of spring, the magic of new buds, leaves and flowers transforms the hills once again with the beauty of woodland life. Delicate cherry blossoms bloom and pheasants give rattling calls from rocky perches, whilst flocks of tits and warblers prey on the first insects that emerge in the new canopy foliage and buzz around the yellow catkins of the early-flowering willows.

The Fragrant Hills park on the outskirts of Beijing is a popular place for visitors. It has excellent botanical gardens and also provides a rare example of northern forests in a near natural state.

ABOVE In autumn, local tourists come to the Fragrant Hills in thousands to see the changing colours and buy souvenir red leaves of maples and aspens.

BELOW The Peking Robin (*Leiothrix lutea*), common in scrubland, is one of the most popular cagebirds in China. In fact, most northern sightings are probably escapes.

ABOVE The deeply fissured bark of the Locust Tree (*Robinia pseudoacacia*) gives a texture and ancient mystery to the temperate forests.

BELOW The Common Fox (*Vulpes vulpes*) is one of China's most widespread mammals but, hunted as a pest and for its fur, is generally very shy.

BELOW RIGHT The Siberian Thrush (*Zoothera sibirica*) breeds in the north-east but migrates through northern China each autumn to its wintering grounds in the south-west. It feeds on berries and insects, mostly on the forest floor.

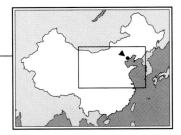

Beidaihe

Although Beidaihe is not recognized as one of the truly wild places in China, this seaside resort, 280 kilometres (174 miles) east of Beijing, is one of the best sites in the world for observing bird migration.

Situated on a triangle of land which protrudes several kilometres into the Gulf of Bohai, between a range of mountains and the coast, the area acts as a funnel for passing migrants. The best times to visit are May and September to November.

Spring migration peaks around the second week of May, when interesting species often include Rufous-bellied Woodpecker, Pechora Pipit, Chinese Penduline Tit and Yellow-browed Bunting. The first half of September is particularly good

for visible migration and there may be over 2,000 Pied Harriers in a day. There is also a good variety of waders at this time and Relict Gull may occur. Later in September and October, Amur Falcon may be seen. At the end of October and beginning of November there is a good chance of seeing Oriental Stork and no fewer than five species of crane – Common, Red-crowned, Siberian, Hooded and White-naped – as well as the impressive Great Bustard.

Many of these birds rarely land but can be seen flying by at close range, making their way to more suitable winter quarters. At Beidaihe, many species previously unrecorded in China have been seen for the first time.

ABOVE The Black-browed Reed Warbler (*Acrocephalus bistrigiceps*) takes the eastern route down the coast of China each autumn.

BELOW The Chestnut-flanked White-eye (*Zosterops erythropleura*) can be seen in flocks mixing with the commoner Japanese White-eye.

ABOVE This female Mugimaki Flycatcher (*Ficedula mugimaki*) is identified by the wing-bar. The brighter male is boldly red-breasted.

BELOW The Yellow-browed Bunting (*Emberiza chrysophrys*) breeds in north-eastern forests but migrates each year to the south.

ABOVE LEFT The Siberian Blue Robin (*Luscinia cyane*) is a regular migrant. It is largely terrestrial and usually seeks dense shade, feeding on small insects among the leaf litter.

ABOVE RIGHT The Relict Gull (*Larus relictus*) is a rare visitor to Beidaihe. The species breeds in Mongolia and on the Ordos plateau reaches of the Yellow river.

RIGHT The pretty Yellow-rumped Flycatcher (*Ficedula zanthopygia*) always stands out with its bright plumage. It keeps to dense thickets close to streams.

BELOW LEFT Boldly barred flanks and undertail coverts identify the shy Baillon's Crake (*Porzana pusilla*) hiding among the coastal scrub.

BELOW RIGHT A beautiful female Amur Falcon (*Falco amurensis*) poses on a telegraph wire at Beidaihe before continuing its southern travels. It can hover like a kestrel.

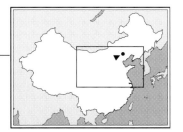

Wulingshan

To the north-east of Beijing City, on the border with Hebei, is another temperate reserve, Wulingshan – larger and wilder than Xiangshan. Here the forests are lusher and less disturbed by visitors. Some ancient huge-trunked trees bear witness to thousands of years of growth and climatic changes. Gnarled, coarse-barked Locust Trees and Ginkgos can still be found here. This is a rare example of the original temperate forests that formerly covered much of this unit.

The ground is moist and mossy, clothed in green ferns and small tree saplings. Nuthatches work through the branches, tree by tree, looking for insects and seeds. They carry acorns to drilling trees where they wedge the nuts in a crack in the bark and hammer away to get at the soft kernels. In summer, gaudy swallowtail butterflies and large fritillaries flutter through the woodland glades and the hammering of the Great Spotted Woodpecker rattles through the forest. The soft cooing of doves gives a peaceful air in total denial of the mass of human development only a few kilometres away.

In autumn, the golden larches and birches change hue and the oak and maple trees put on a show of red leaves of startling brightness, matched by the scarlet *Russula* toadstools that push their way up among the vegetation of the forest floor. Ants scurry to hoard enough food to last them through the winter and harsh-calling Eurasian Jays pluck the last acorns from the leafy oak trees. The treetops are alive with the twittering of passing flocks of white-eyes feasting on autumn insects before moving on migration to southern China.

BELOW Wulingshan Nature Reserve, located in valleys that cluster around Mount Wulingshan, is quite hard to find despite being relatively close to the national capital. A view southwards from the mountain's upper slopes clearly shows the transition from forest within the reserve to denuded hills outside.
OPPOSITE PAGE, ABOVE AND BELOW The reserve's forests are mixed broadleaf-coniferous, some of the common trees including birch, oak, maple and larch, which in autumn are a blaze of colour. A few wild Ginkgo trees are also found.

ABOVE The Locust Tree (*Robinia pseudoacacia*) is rare in China, although there are several species in North America. It forms part of the great mixed broadleaf forests, once so rich but now so depleted.

ABOVE Larches (*Larix*) are among the common conifers of northern China. Unlike most other conifers, which are evergreen, they lose their leaves in winter.

BELOW Maples (*Acer*) are among the dominant genera of the temperate broadleaf forests. They also provide some of the brightest colours in autumn.

BELOW Rhesus Macaques (*Macaca mulatta*) are the most northerly primates in China, reaching Shaanxi and Hebei provinces up to the same level as Beijing.

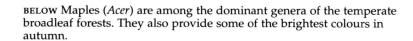

Focus on
North-west China

The north-west is a huge and relatively barren unit of China with a sparse population. The total region consists of some 3 million square kilometres (1¼ million square miles) comprising, in the east, the great Inner Mongolian plain, parts of the Gobi desert, and the Ordos plateau, together with their associated mountain ranges – Helan, Yinshan and the southern extension of the Da Hinggan range. Elevations on these plateaux are generally between 1,000 and 1,500 metres (3,280–4,900 feet). In the west this unit includes the Alashan plateau, two great arid desert depressions – the Junggar and Tarim basins – and two major mountain ranges, the Altai on the north-west borders with Russia and the Tianshan (Mountains of Heaven) which run east-west dividing the deserts. Elevations in the two basins range between 500 and 1,000 metres (1,600–3,280 feet), and in the small Turpan basin fall as low as 155 metres (500 feet) below sea level,

the lowest point in mainland Asia. The highest peaks of the Tianshan range are covered in snow and reach up to 7,435 metres (24,393 feet) at the summit of Mount Tomul.

The only significant forests of the unit grow on the Altai and Tianshan mountain ranges. On Tianshan, montane grasslands gradually give way to coniferous forests of spruce (*Picea*) with alpine grassy meadows and groves of round-leaved birch. In the Altai mountains firs and larches dominate the conifer zone.

Climate in the region varies with altitude and latitude, with cool-temperate weather in the north and warm-temperate conditions predominating further south. Aridity increases westwards, with rainfall from 100 to 200 millimetres (4–8 inches) giving desert conditions, compared with up to 500 millimetres (20 inches) in the east, where lush grasslands provide superb pasture for the nomadic herdsmen and their sheep, horses and

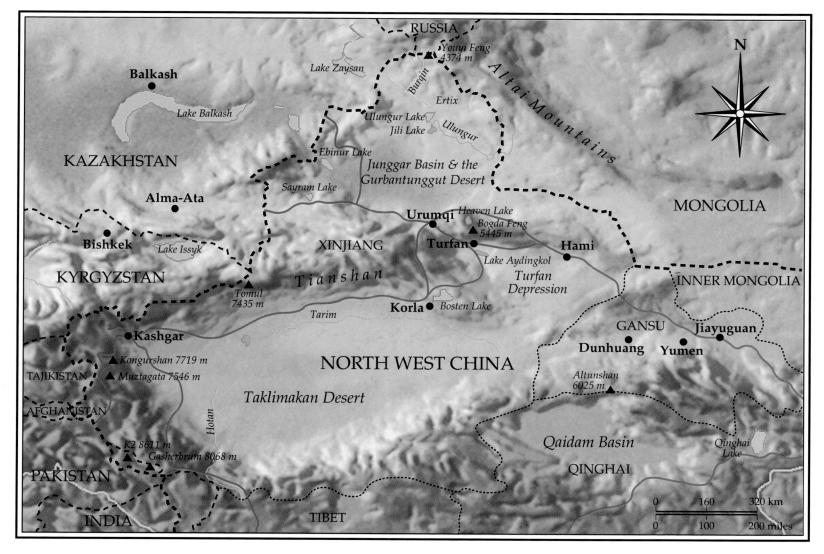

ABOVE Receiving very little rain, much of China's north-west is desert. The region includes the Taklimakan, the world's second largest desert of shifting sand.

RIGHT In areas where precipitation is higher forests abound, as in Xinjiang's Kanas Lake Nature Reserve in the Altai mountains bordering Russia and Mongolia.

cows. In former times, the climate was probably less arid and the pockets of riparian forests comprising poplars and elms suggest there was once a more parklike savannah vegetation. Today, however, as a result of continuing desiccation a more xerophilous *Artemisia* scrub vegetation dominates much of the unit.

The Altai mountains extend for over 2,000 kilometres (1,200 miles) along the Mongolian-China and China-Russia border, and thence deep into Russia. The word *altai* is Mongolian for 'gold' – during the Tang dynasty this was a rich gold-mining area. Only about 500 kilometres (300 miles) of the central southern slopes lie within China. These rise to an elevation along the ridge of just over 3,000 metres (9,800 feet) with the highest peak of Youyi or 'Friendship' at 4,374 metres (14,350 feet). The summits of Altai are covered in continual snow and there are many well-formed glaciers. Winter arrives early with the first frosts reaching the upper slopes by early August. Snow and rainfall are quite high, so many small rivers drain down from the mountains bringing valuable water into the desert regions immediately to the south in the Junggar basin.

The Tianshan are one of China's major mountain ranges. The western end of the range lies within Kyrgyzstan but more than 20 parallel ranges that make up the mountains stretch for 1,700 kilometres (1,000 miles) across Xinjiang province, separating the Junggar basin to the north from the Tarim basin to the south. The mountains have an area of over 240,000 square kilometres (90,000 square miles), with most ridges between 3,000 and 4,000 metres (9,800–13,000 feet).

Little moisture drifts north off the Tarim and the southern faces of Tianshan therefore have a dry mountain steppe vegetation but, on the north, moisture from the Arctic does reach Tianshan to support a denser vegetation and extensive forests of Schrenck's Spruce (*Picea schrenckiana*). In winter these forests are clothed in a layer of snow which melts in spring to swell the lakes and streams of these beautiful mountains. There are over 6,000 different glaciers on the upper slopes.

The Altai and Tianshan ranges resemble north-east China in their fauna, with wolves, foxes, sable, wolverine, wild pig, roedeer and chipmunks as characteristic mammals. Altai has been invaded by Red Squirrels (*Sciurus vulgaris*) from Russia, and is also notable for protecting the only population of the Beaver (*Castor fiber*) in China. Beavers live in the upper rivers and create dams of logs and gravel to block streams, creating deep pools in which they make their nests. They feed on tree bark and aquatic vegetation and store up food in their semi-submerged dens to keep them through the frigid winter months. Many of these

mammals are trapped for their fur; the sable, in particular, is much coveted but foxes and squirrels are also popular.

The lower grasslands and semi-deserts have a completely different fauna. The variety of rodent mammals is significant, with many species of voles and marmots and rodent-like picas. Large mole-like rodents called Zokors (*Myospalax fontanieri*) burrow under the grassland, emerging periodically to throw out the excess earth in loose piles. Zokors eat worms and have a voracious appetite. There are millions of them in the grasslands

and Chinese pharmacists have started to use their bones as a substitute for tigers' bones in traditional medicines. It would be a good thing if the practice became widespread and these plentiful animals could take the pressure off one of the most endangered mammals in Asia.

Large mammals of the grasslands include wild camels, graceful gazelles and the fast-running Wild Ass (*Equus hemionus*). The beautiful Przewalski's horses (*Equus przewalskii*) with their chunky form, so reminiscent of the Tang dynasty figurines, have

ABOVE LEFT Przewalski's Horse (*Equus przewalskii*) is extinct in the wild. Efforts are being made to reintroduce the species to China.
ABOVE RIGHT A few Bactrian Camels (*Camelus bactrianus*) still occur wild in the deserts of Xinjiang and Inner Mongolia.
LEFT The Flaming mountains, near the Turpan oasis, Xinjiang, are aptly named both for their reddish colour and the searing temperatures reached during the daytime.

become extinct in the wild but captive animals are currently being reintroduced and some wild herds will soon be re-established in northern Gansu and Xinjiang.

Animals of the desert face very harsh conditions and can survive only with the aid of some special adaptations to the arid environment. The larger species cannot find shelter and are exposed to the full heat of the day. Camels, wild asses, gazelles and desert cats find sleep the best defence and are active only at night or in the cool of dawn and dusk. Smaller animals, such as rodents and reptiles, can escape the mid-day sun by burrowing to the cool and dark subterranean world, emerging only at night to go about their business.

Two groups of rodents – the jerboas and gerbils – are particularly successful in the desert and these both have opted for a hopping mode of locomotion. In jerboas the hind-legs are as much as five times the length of the forelegs and they bound in long kangaroo leaps, keeping their balance with a long flexible tail. Several different species occur in different desert regions and habitats. Some have large batlike ears which help them to hear the danger signs of any approaching snake or fox. There are several species of gerbils, too. This group is less extreme in form than the jerboas, with hind-legs only twice as long as the forelegs. They live in complex burrow systems, breeding up to high densities whenever rains trigger a flush of plants and seeds. When food becomes scarce, mortality knocks back the numbers and some animals migrate in search of new habitat. These animals can extract water from even the driest foods.

Goitred Gazelle (*Gazella subgutturosa*) and Przewalski's Gazelle (*Procapra przewalskii*) live in small herds, wandering across the edges of the deserts wherever they can find plants to eat. They are confined largely to the stony deserts and both are declining as a result of hunting. Camels, too, have become very rare now. Although they are valued for domestication, there are gangs hunting the wild animals for meat, hides and bones. All the camel's famous hardiness and ability to live for weeks without food and water give it no protection against the rifle and jeep.

Desert reptiles include the endemic Horsfield's Tortoise (*Testudo horsfieldi*), several agamid lizards, geckos and a few snakes. The tortoise is a herbivore but the others feed on grasshoppers or, in the case of snakes such as the Sand Boa (*Eryx miliaris*), on small rodents. The boa lives in the burrows of its chief prey or can bury itself in soft sand.

Desert birds cannot reduce heat by sweating but can cool themselves in the wind by puffing out their feathers and also by panting. Typical desert species include predatory falcons and hawks, bustards, rollers, nightjars, larks, shrikes, ground-jays and wheatears. One of the most intriguing desert birds is the sandgrouse. These birds fly enormous distances each day to find water, then return to the desert to feed and breed. Male sandgrouse even carry water back in their beaks to pass on to their chicks in the nest which is made in a shallow scrape in the gravel. At certain times of the year, the sandgrouse form large migrating flocks and fly south to the cooler and moister areas.

With little natural cover, desert animals have to rely on camouflage to conceal themselves from predators. Most are sandy coloured to blend into their background and many have counter-shading with pale underparts and darker upperparts to disguise the pattern of shadows and flatten their apparent shape. Many also have blotched or patchy patterns to break up or disrupt their outline and make them less conspicuous sitting motionless in the open. The geckos and some agamids can even change their colour to suit their background like chameleons.

Desert conditions place extreme restrictions on the growth and survival of plants. These include shortage of moisture,

The calm waters of Western Heaven lake, backed by the snowy peak of Bogda Feng in the Tianshan range, are one of the classic views of north-west China.

intense temperature fluctuations, fast desiccation in wind and sunshine and often saline soils caused by evaporation of surface water. Few species can tolerate such conditions and those that are able to live in such inhospitable terrain possess unusual adaptations.

For a start, desert plants tend to have long deep roots to reach moisture that remains well below the surface and where the salt is less concentrated by evaporation. Many species have water-storing tubers which swell up in the moister times of year and sustain the plants through the driest months.

The design of desert plant leaves is adapted to minimize moisture losses through transpiration. They tend to be small and elongated or needle-like in shape, or deeply divided, as in the common desert wormwoods (*Artemisia*). Some plants have reduced numbers of stomata, others have thickened cuticle or a dense covering of fine hairs on leaves and stems, such as in the Orache (*Atriplex*). The common desert shrubs *Tamarix* and *Reaumuria* are able to secrete excess salt through their stomata, have waxy cuticles to reduce water loss and lose their leaves and become dormant in periods of extreme drought. The Fat Hen Goosefoot (*Chenopodium album*) has thick linear leaves with a sticky sap that can absorb a lot of soluble salts. Grasses such as desert fescue (*Festuca*) and needle-grass (*Stipa*) roll up their leaves through the heat of the day to reduce water loss, whilst the Leafless Anabasis (*Anabasis aphylla*) has gone further and does most of its photosynthesis through its stems.

Different plants are adapted to specific parts of the desert environment. Thus the stony tracks of dried river beds are dominated by a sparse cover of *Anabasis*, *Reaumuria*, Kaschgar Ephedra (*Ephedra przewalskii*) and Siberian Nitraria (*Nitraria sphaerocarpa*), which can all root into hard cracks. Other species can cope with the problems of soft shifting sands. Here, one can find Mongolian Calligonum (*Calligonum mongolicum*), Squarrose Agriophyllum (*Agriophyllum arenarium*) and milkvetches (*Astragalus*). Firmer dunes are stabilized by saxauls (*Haloxylon*) and Beancaper (*Zygophyllum fabago*). Salty areas are dominated by *Kalidium* species, which hold so much salt that they are unpalatable to herbivores until dead and dried up and the salt has been leached away.

Many ephemeral annual plants remain dormant in the desert sand and gravel as seeds, waiting for those rare rainstorms that will trigger germination and allow a brief blooming before another barren wait of several years.

Below the deserts there are often artesian water-fields. Where these come near to the surface, oases occur. A sudden patch of poplar trees (*Populus*) or even elms (*Ulmus pumila*) shine shockingly green in the arid landscape of red rock and yellow sands. Wells have been dug to reach the underground water and these oases are important centres of human life – small villages in the desert or temporary stopping points for the nomadic camel and goat herders.

Despite the low human population density in north-west China, the rate of population increase is very high and concurrent pressures such as agricultural development, hunting and trapping grow ever more severe. Falcons are trapped in large numbers to satisfy the growing demand in Pakistan and Arab countries, where falconers may pay thousands of dollars for a single bird. All fur and meat-bearing animals remain at risk. Even more of a problem is the fencing of the grasslands. Both wild and domestic animals used to roam with the seasons over the great plains and plateaux of the unit but the recent tendency to enclose these lands for ranching or farming is proving disastrous to many species.

Another threat comes from the sinking of water bore-holes. These create local oases which lead to local overstocking and overgrazing, whilst over the wider area the water table becomes lowered and the desertification process continues. One only has to look to the huge deserts of India and northern Africa to see the kind of desertification that can occur through centuries of overgrazing and disregard of the natural replenishment processes of the ecosystems upon which herdsmen depend. Another feature of overgrazing is that the desirable species become fewer and smaller but toxic, spiny and unusable plants start to predominate and become invasive.

Altai Mountains

Only a small part of the Altai range is protected, and increasing logging and grazing are threatening the rather distinctive fauna and flora of these interesting mountains, which form China's north-west ramparts. Vegetation is dense in the north-western parts but sparse in drier areas to the south and east, and altitudinal zones are well defined.

Immediately below the permanent snowline there is a narrow belt of alpine cushion vegetation, with many lichens and mosses clinging to the rocky ground. Below this is a broader zone of alpine meadows and scrub. Round-leaved birch stands occur in the western parts and in the east there are meadows of *Kobresia* and *Carex*, where picas and marmots tease the Saker Falcons. A zone of subalpine grasslands occurs between 2,300 and 2,600 metres (7,500–8,500 feet), dominated by *Poa* and *Festuca* grasses and providing excellent grazing for deer and rodents. However, most of this zone is now used as pasture for domestic herds.

Conifer forests form dense cover between 1,800 and 2,300 metres (5,900–7,500 feet), depending on aspect and aridity. Siberian Spruce and Siberian Larch are the main species and this region of about a million hectares (2,500,000 acres) is the most important source of timber in north-west China. Beavers still inhabit the eastern rivers of this zone and roedeer winter on the forests before moving to higher altitudes to feed in the summer. Below the conifer zone is a mosaic of shrub-steppes in drier and lower parts. These provide limited spring grazing but become very dry in the summer months.

The Kanas river thunders downhill through coniferous forests close to the southern limit of Kanas Lake Nature Reserve in the Altaishan. The only part of the Altai range within China to be protected, the reserve occupies the northernmost apex of Xinjiang and is bordered by Kazakhstan, Russia and Mongolia.

RIGHT The Kanas river snakes its way through spruce and larch forest. Its milky coloration is due to suspended rock particles, the result of erosion by glaciers streaming down Friendship Peak, the 4,374-metre (14,350-foot) mountain that is the highest in the Kanas reserve. The mountain's name is derived from the fact that it straddles the Sino-Russian border.

BELOW LEFT A Siberian Spruce (*Picea obovata*), one of the main species of the Altaishan's conifer forests, stands above the waters of Kanas lake.

BELOW RIGHT Dense forest in the Altaishan, showing some conifers but a large proportion of birch and poplar, the latter starting to turn an autumnal gold. Winter comes early to this remote northern region, the deciduous trees starting to take on autumn colours by late August, coincident with the first light snow showers.

ABOVE The Common Swallowtail butterfly (*Papilio machaon*) ranges from Europe to the Himalayas and Siberia. Despite the fragility of the Western race, this is a hardy insect in northern China, where it feeds on several plants of the wild carrot family.

·BELOW LEFT Cones and the tightly bunched groups of leaves of Siberian Larch (*Larix sibirica*), the second of the two main species of conifer found in the Altai range.

BELOW CENTRE The forests of the Altai are very moist. Mushrooms (*Lactarius*) squeeze through to join the dense temperate ground herbage.

BELOW RIGHT The dark and damp conditions beneath the coniferous tree cover also create an ideal environment for the growth of moss and lichen, resulting in lush beds over rocks and logs throughout much of the forest.

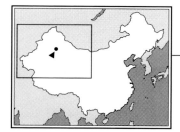

Mountains of Heaven

The Mountains of Heaven (Tianshan) form a highly complex range which has been folded many times. Between its many ridges, deep valleys, grabens and intermontane basins lie trapped, holding lakes and more moisture than the outer faces, and supporting a surprisingly rich vegetation. Here, also, more than 200 different rivers have their source. Perhaps most beautiful of Tianshan's hidden waterways is the clear blue Western Heaven lake surrounded by stunning alpine scenery of stately spruce forests, steep mountains and open meadows of pretty flowers.

Like the Altai, the range shows sharp altitudinal zonation of vegetation. The main spruce forests are confined to a belt between 2,700 and 3,700 metres (8,800–12,000 feet). Above this, forests give way to alpine meadows of *Carex* and *Alchemilla* with grasses such as *Poa pratensis* and *Bronmus inermis*. This zone provides excellent pasture for the Kazakh herdsmen. From here to the treeline at almost 4,000 metres (13,120 feet) alpine cushion vegetation persists until bare rock and ice dominate the high landscape.

Below the forests are arid montane steppes dominated by *Artemisia* shrubs and grasses of *Festuca* and *Stipa* species. Artificial tree belts of poplars have been planted along the foothills to contain the deserts.

Western Heaven lake in the Tianshan, a mountain range that divides Xinjiang. It is one of the few green areas in the whole of the north-west, the high peaks catching what little moisture there is, resulting in permanent snow, cascading rivers, lakes, lush pastures and dense forests.

Known as Tianchi in Chinese, Western Heaven lake is probably one of the most beautiful and famous sights in the whole of China's north-west. Though part of the lake shore is heavily touristed, most of the region is still quiet and undisturbed, its forests a valuable nature reserve, the grasslands providing pasture for local Kazakh herdsmen.

ABOVE Classic Tianshan landscape consisting of mountains covered with rocky outcrops and either dense coniferous forest or open pastures, the latter usually on south-facing slopes.

OPPOSITE PAGE, ABOVE The snow-capped peak is Bogda Feng, the highest mountain in the easternmost part of the Tianshan range.

LEFT AND OPPOSITE PAGE, BELOW The forest often consists of almost pure stands of Schrenck's Spruce (*Picea schrenckiana*), a tree characterized by its tall and very narrow spire-like shape.

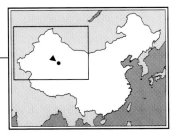

Dunhuang

Dunhuang, in the far north-west of Gansu province, is in one of the driest places in China, an extensive oasis that was a major staging post for the Silk Road. It was, and still is, a haven from the vast deserts of the north-west for those on the road. Sandy and stony desert crowds in close all around Dunhuang, coming right up to the last ranks of protective poplar trees and stretching away for hundreds of kilometres in all directions.

Dunhuang provides one of the best gateways to some of China's most barren desert scenery. Huge dunes nearly 300 metres (1,000 feet) high – the Whispering Sands – come right to the edge of the oasis, giving a more than adequate foretaste of the shifting sands of the inaccessible Taklimakan desert further to the west. Where there are no dunes, gravelly desert stretches away towards lines of desolate hills, eroded and much-gullied, coloured grey, purple and pink according to the types of rock exposed. There is no shade and the heat of the sun is intense.

Not much survives in the burning sands, yet lizards are quite common. These include the nocturnal Ring-tailed Gecko, while early in the morning or shortly before sunset one may see the strange toad-headed lizard *Phrynocephalus albolineautus*, a member of the long-legged agamid family. Plants are few and far between, though a few grasses manage to get a foothold even on the dunes, while clusters of xerophytic bushes may be found in the occasional dry, stony river bed.

Looking from the tops of some of the dunes of the Whispering Sands, it is easy to see how these huge mountains of sand tower menacingly above the outer edges of the Dunhuang oasis, the distant trees marking the limits of irrigation and fertility.

RIGHT The interplay of light thrown by the setting sun onto the barren dunes around Dunhuang creates a striking geometric pattern in simple colours. The diagonal line in the foreground illustrates well just how razor-sharp are many of the dunes' uppermost edges, a steep but climbable windward side switching sharply into an almost vertical leeward wall.

BELOW Even after sunset the shapes and colours of the dunes retain their interest. A full moon rises above a dune that is still lit by the last of the light of a fiery dusk, its smooth slope taking on an attractive pink hue.

LEFT The toad-headed *Phrynocephalus forsythi* is an agamid lizard that braves the desert sands in search of insects, burying itself underground for shelter during the heat of the day.

LEFT The Chukar (*Alectoris chukar*) is a most hardy bird living in some of the driest, coldest and least hospitable parts of northern China. It lives in small coveys and sleeps through the day.

BELOW LEFT A dry, stony riverbed in the desert beyond Dunhuang provides one of the few spots where hardy bushes can find just enough moisture and anchorage to survive.

BELOW RIGHT Incredibly, a few plants are able to survive on the dunes, anchoring themselves deep into the sand and reducing both leaf size and number in order to minimize water loss.

Focus on
The Tibetan Plateau
and Himalayas

This is another enormous region with a total area of over 2.5 million square kilometres (1 million square miles), or one quarter of the entire country. It is also the most desolate and wild part of China with less than one per cent of the nation's population and agricultural area.

The Tibetan Plateau

Much of the Tibetan (Xizang-Qinghai) plateau was formerly an ancient sea called Tethys. However, with the impact of the Indian subcontinent into the belly of mainland Asia some 30 million years ago, the buckling and rising processes began, that are still continuing today, forcing up the world's largest high-altitude plateau and, at its southern fringe, the world's highest and one of its youngest mountain ranges – the Himalayas.

At first view, the plateau is uninviting – an endless waste of bleak and barren landscape, visited by extreme cold and biting winds. But wait until your head clears of the high-altitude dizziness, wait until your face gets used to the numbing cold and

your eyes stop watering in the wind, and take a longer look. You will find a world of enormous variety and beauty, an intricate mosaic of snow-capped mountain ranges, rolling hills, stony deserts, reedy marshes and sparkling lakes.

The entire plateau area exerts great climatic effect on the surrounding land due to its high pressure, vast size and general aridity. Few clouds can rise up above the plateau, giving it the sunniest weather in the whole of China but also extremely dry conditions. It is thus a cold desert, quite different in nature from the warm sandy deserts of the north-west China unit. There are huge diurnal and seasonal variations in temperature and the constant thawing and freezing lead to high levels of erosion and weathering. Cooler regions in the north and centre of the plateau are permanently frozen below the ground surface.

The plateau is pocked by thousands of small lakes. Where isolated lakes fed by glaciers have been exposed to thousands of years of sunshine and evaporation they become salty, some turning into pans of solid crystal salt. Others are quite fresh and support a rich water vegetation. There are red lakes, coloured by

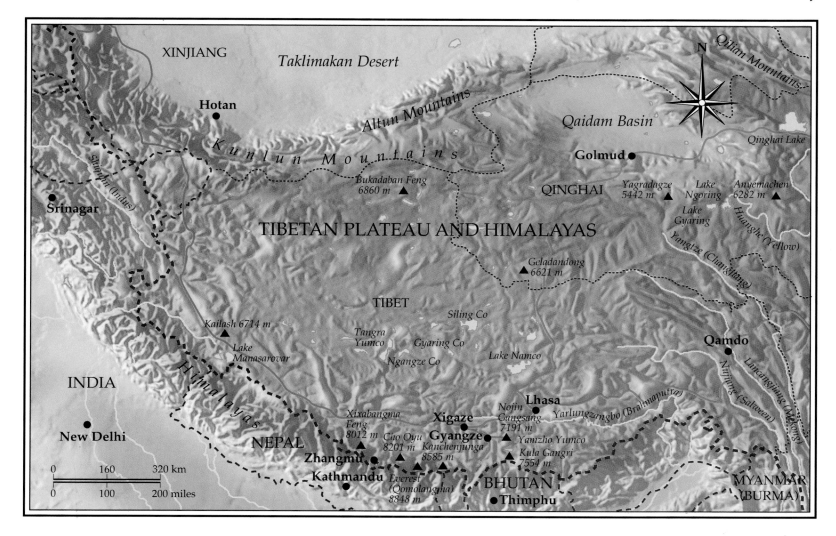

Huge and desolate, the Tibetan plateau and Himalayan region is far from lacking in interesting species. TOP LEFT The Common Muntjac (*Muntiacus muntjak*). TOP RIGHT Blue Sheep (*Pseudois nayaur*) live in large herds on the high alpine grasslands. ABOVE LEFT Himalayan Monal (*Lophophorus impejanus*), a spectacular pheasant of the upper conifer forest and alpine scrub. ABOVE CENTRE The Stripe-throated Yuhina (*Yuhina gularis*), common in Himalayan conifer forests. ABOVE RIGHT A brightly coloured shield bug (suborder Heteroptera) warns of its offensive taste. FAR LEFT A tiny hart's-tongue fern (*Phyllitis*) in the lush Zhangmu Kou'an gorge. LEFT Tender fir shoots (*Abies*) in the dense conifer zone.

TOP LEFT *Sassaurea* has white paperlike bracts around the flowerheads that make the plant stand out against the green hillside. TOP CENTRE Epiphytic orchids (*Pleione*) cluster on mossy branches and on the floor of forest glades. TOP RIGHT The delicate shape and rich hue of a wild iris (*Iris scariosa*). ABOVE LEFT A yellow poppywort (*Mecanopsis paniculata*) grows tall in a sheltered spot. ABOVE CENTRE An alpine geranium shines in morning sunlight. ABOVE RIGHT Louseworts (*Pedicularis anas*) are a familiar sight, glowing pink or mauve among the alpine flora. RIGHT Cowslips (*Primula*) are very common in the region. There are many species adapted to different altitudes and conditions. FAR RIGHT Wild roses (*Rosa webbiana*) are abundant in the Himalayas.

The conical peak of 7,191-metre (23,592-foot) Mount Nojin Gangsang is a major feature of the Tibetan landscape south-west of Lhasa, dominating the southern end of Yamdrok Tso lake, when seen from the Kampa La pass and the road between the Tibetan capital and the town of Gyantse.

microscopic algae, and others almost boiling hot, fed by the hot springs that bubble up from the many cracks in this tectonically dynamic part of the earth's crust.

The great Kunlun range of mountains forms the northern border of the Tibetan plateau. It is, in fact, several ranges made up from west to east by the West Kunlun, Qilian and Altun and includes the curious Qaidam basin, which is a graben depression in the northern part of the plateau. These mountains are the driest region in the whole of China. Parallel mountain ranges of southern Tibet block any maritime air masses that may carry moisture up onto the plateau from the south, whilst the successive ranges of Altai, Tianshan and the northern Kunlun stand as an effective barrier to any moisture from the Arctic.

The vegetation of the plateau region is sometimes poor in species and ground cover but quite distinctive, with a high level of species endemism. In the north-west the flora is derived from the panarctic flora with which there were connections during the last Ice Age. To the south-east the vegetation gets richer with gradual intermixture from the Indo-Malayan and Himalayan floras. Many alpine genera occur and the important group of rhododendrons and azaleas has its distribution centre here.

For most of the plateau there are only four months of summer growing season. The *Stipa* and *Kobresia* grasslands turn lush to feed the wild herbivores and domestic stock alike, and the alpine meadows are full of flowers – masses and masses of them, hundreds of species packed together on a single piece of land. Louseworts, asters and vetches splash pink and mauve on the scene. *Mecanopsis* poppies add bright blue and sulphur yellow, irises and gentians regal purple, and the *Corydalis* contributes a neon shade of pale blue that is almost unbelievable. Delicate primroses form stalked pompoms of varied hues and beautiful lilies and fritillaries lend elegance to the uninhibited masses of colour. This is the supreme time of beauty on the plateau. Stop to admire it while it lasts, listen to the buzz of insect wings and the song of the lark high in the sky, and relish the vast wildness of it all.

Here, too, are countless mysteries. The red spikes of the *Polygonum* may not look special amid so many other gorgeous plants but this species is unusual in sprouting its seeds while they are still on the flower. Moreover, the seeds are so rich in starch that this is an important food plant for animals and even humans on the plateau. The delicate *Drosera* that grows around swampy ground has curious hairs on its leaves that exude a sticky sweet liquid to first attract then trap visiting flies. The leaves respond by curling over the insects and digestive fluids flow to extract the nutrition from their captives.

Even more bizarre is the oddity known to Chinese medical

practitioners as *dongchong xiacao*, 'winter larva, summer grass'. This is in fact the larva of an insect, *Hepialus virescens*, that hibernates under the soil but is sometimes attacked by a fungus, *Cordyceps sinensis*, which grows a tall grass-like fruiting stem that sticks out of the ground. Farmers recognize the stems and dig up the dead larvae which are sold as a valuable bitter-tasting medicine, said to strengthen the lungs and enhance virility.

Many of the other plants of the plateau have medicinal properties. Roots of the orchid *Gastrodia elata* are collected as a remedy for headaches and blurred vision. Bulbs of fritillaries are used to cure colds. The leaves and seeds of the blue-flowering bush *Sophora moorcroftiana* are used as a diuretic and antipyretic and even act as an insecticide. The berries of the spiny *Berberis* bushes are used to cure dysentery.

In parts of the plateau that are not fed by melting glaciers, and are dependent for moisture on the cold mists of winter and rare snow showers, the land is covered in dry stones in every direction and few plants can gain a foothold. Those that do have developed several tricks for survival in the harsh conditions. Most keep their heads down and stay close to the ground to escape the bitter and drying winds but many cling or lie flat to an unusual degree. *Myricayia prostrata* grows laterally rather than standing up in the wind. The leaves of *Lamiphlomis rotata*, *Microula tibetica* and *Oreosolen wattii* spread broad and flat over the ground. Here they protect the soil beneath them from drying out, suppress competition by smothering neighbouring plants

Yamdrok Tso, at an altitude of about 4,400 metres (14,400 feet), is one of the largest of Tibet's many salt lakes, the last remnants of the Tethys Sea. A number of shoreline marshy grasslands provide valuable habitat for migratory waders, ducks and geese.

and make themselves difficult for herbivores to eat. Other alpine plants have developed a tight dense cushion form, with many tiny leaves and flowers, which helps to maximize heat absorption from sunshine whilst minimizing the rate of heat loss at night. The dense structure and compact surface also reduce evaporation and wind resistance. So successful is this adaptation to environment that literally hundreds of species of many different families and genera have independently adopted the form and whole areas of the plateau are covered by what is known as alpine cushion vegetation.

Bushes of *Cotoneaster* have flattened branches and tiny leaves in adaptation to clinging tight to rocks to reduce wind resistance in the harsh alpine zone. Their dense red berries provide welcome food for the scrub birds that also act as seed disperal agents for these pretty bushes. Several species of *Cotoneaster* have been reared for horticultural purposes because of their colourful berries and suitability for training against walls in distant western gardens.

In all, over 5,700 species of higher plants are recorded in Tibet, and if algae, mosses, fungi and lichens were added the total would exceed 10,000.

The fauna of the plateau area is characterized by generally low densities and great hardiness. The great shaggy Wild Yaks (*Bos grunniens*), which have been domesticated by Tibetan herdsmen for centuries, have their home here, and a high-altitude sub-species of wild ass (*Equus hemionus kiang*) frequents the desert regions where competition for food is minimal. Herds of Tibetan Antelope (*Pantholops hodgsoni*) and Tibetan Gazelle (*Procapra picticaudata*) feed on the sparsely vegetated open slopes and several species of wild sheep graze the mountain sides. Commonest is the Blue Sheep or Bharal (*Pseudois nayaur*) which lives in large herds of a hundred or more animals. The males have widely separating, gently spiralling horns, which differ markedly from the hefty recurved horns of the much rarer Argali or Marco Polo Sheep (*Ovis ammon*).

The Rufous-vented Yuhina (*Yuhina occipitalis*) is a common flocking bird at moderate altitudes in southern Tibet and south-west China. It often associates with tits and flycatchers to make mixed flock bird 'waves'.

ABOVE The Lhasa, or Kyichu, river running downstream of Lhasa a few kilometres before its confluence with the Yarlungzangbo, the upper Brahmaputra.

OPPOSITE PAGE A salt pan near Old Tingri in the far south of Tibet. Many of Tibet's wetlands are highly saline, and salt pans are not uncommon.

Smaller mammals include the robust marmots – giant ground squirrels that live in rabbit-like burrows in small colonies. These sandy brown rodents feed on grasses and tender herbs. They walk cautiously further and further away from the safety of their holes but stand tall for any sign of their many enemies. A fox stalking through the cover of rocks and bushes, an eagle diving lazily out of the blue sky, using the dazzling sun to hide his attack – these are the images that keep the marmot alert or send it scurrying back to its burrow.

Picas are small brown rabbits with short rounded ears and beady eyes. Several species live on the plateau adapted to different habitats or confined to particular mountain ranges. At an even smaller level there are many voles (*Alticola*) burrowing under shallow alpine soil or, in winter, just under the snow but above the soil.

Preying on these variously sized herbivores is a long list of carnivores – Brown Bear (*Ursus arctos*), Wolf (*Canis lupus*), foxes (*Vulpes vulpes*), Lynx (*Felis lynx*), Pallas's Cat (*F. manul*) and several weasels. The most beautiful and mysterious of all is the elusive Snow Leopard (*Uncia uncia*) whose soft grey coat is so admired by the fur industry and blends so marvellously with the rocks and lichens of the animal's range. The Snow Leopard follows the herds of Blue Sheep, day by day and week by week, cutting off the sick, the young and the unlucky.

Eagles, buzzards, vultures, hawks, falcons and owls add their skills to the already formidable array of mammalian predators. For this is perfect raptor country – wide open spaces, shortage of cover and plenty of environmental hardship to distract and blunt the senses of the prey. High-altitude species include the Golden Eagle (*Aquila chrysaetos*), Himalayan Griffon (*Gyps himalayensis*) and the amazing Lammergeier or Bearded Vulture (*Gypaetus*

barbatus). The last has the fascinating habit of carrying the bones of dead animals high into the sky before dropping them onto rocks to smash them up in order to feed on the marrow inside.

Other wonderful birds have made the harsh plateau their own. The noisy Tibetan Snowcock (*Tetraogallus tibetanus*) and Snow Partridge (*Lerwa lerwa*) strut among the alpine screes, and Snow Pigeons (*Columba leuconota*) glide and play in the wind, showing off their beautiful white plumage. The incredibly blue Grandala (*Grandala coelicolor*) perches on a rock to watch his domain, while flocks of mountain-finches (*Leucosticte*) and rosefinches (*Carpodacus*) feed among the scrub and grasslands. The clear-running streams are patrolled by White-capped Water Redstarts (*Chaimarrornis leucocephalus*) and Blue Whistling-Thrushes (*Myiophonus caeruleus*). In the summer months, swans, Bar-headed Geese (*Anser indicus*) and Black-necked Cranes (*Grus nigricollis*) breed in the marshes and lakesides, but in winter these areas are frozen up and the birds fly to milder wintering grounds to the south and east.

ABOVE In early summer a thaw sets in among snow on the Gyatso La pass, at 5,220 metres (17,126 feet) the highest pass crossed by the 1,000–kilometre (620–mile) Friendship Highway linking Lhasa and Kathmandu.
LEFT An abrupt end for a glacier streaming down the slopes of Mount Nojin Gangsang. Although the air on the Tibetan plateau is extremely dry, storms are not uncommon, leaving the highest peaks with a permanent covering of snow.

OPPOSITE PAGE, ABOVE The mysterious Snow Leopard (*Uncia uncia*) is the most beautiful of Asia's wild cats. Its fluffy grey fur blends marvellously into the rocky landscape and few people ever see this great hunter.
OPPOSITE PAGE, BELOW Wild Yaks (*Bos grunniens*) are confined to the high plateau areas above 4,000 metres (13,120 feet). Through the winter they feed on dried grasses which they can find under the shallow snow. In summer, they fatten up on new growth.

ABOVE The landscape of southern Tibet as seen from the Younong La pass. Vegetation is sparse, and the desolation of the Tibetan plateau is extreme, something that should not be underestimated by any traveller.

BELOW The large, dark Cineraceous Vulture (*Aegypius monachus*) lives over much of eastern Tibet. Vultures are ideal raptors in very open extensive landscapes as they can fly great distances on thermals and spot dead or dying animals from a great height.

The Ruddy Shelduck (*Tadorna ferruginea*) breeds on the many freshwater and saline lakes that dot the Tibetan plateau. It is a handsome and hardy bird that is found up to 4,500 metres (14,800 feet) and can be quite numerous.

The never-ending snow-capped mountain ranges of eastern Tibet. Consisting of several north-south lines of mountains that are eastern extensions of the Himalayas, this remains one of the remotest regions on earth, parts of it barely explored even by the Chinese.

Despite the low human densities, man has a large impact on the local fauna. There is little cover and powerful rifles with telescopic sights bring everything within the hunters' range. Great numbers of the larger animals are killed each year.

Another major problem is the shortage of fuel, which has led to villagers cutting more and more of the sparse vegetation for the stove and degrading the cover available for wildlife. Traditional pastoralists burn animal dung but the immigrant Chinese prefer wood and the destruction of cover leads to ethnic as well as ecological tensions.

The Himalayas

In China, the Himalayas are called the 'roof of the world'. The range extends for some 5,000 kilometres (3,000 miles) along the southern border of Tibet before fanning out into a number of smaller extension ranges in western Yunnan. Many peaks rise to over 7,000 metres (23,000 feet), with the world's highest peak, Mount Everest or Qomolangma, reaching a mighty 8,848 metres (29,029 feet).

Biologically the highest altitudes are a barren zone of bare rock and permanent ice and snow but at lower altitudes the Himalayas are regarded as one of the world's ten biodiversity 'hot spots', characterized by an extraordinary variety of plants, birds and other groups.

To the north the Himalayas descend only as far as the Tibetan plateau and the deep valley of the Yarlungzangbo river, which turns south through a gap in the eastern Himalayas to become the great Brahmaputra. The north faces of the range are cold and rather dry, supporting only alpine meadows and scrub. However, the southern faces catch and hold the warm moist weather from tropical Asia. This is where the real richness of the Himalayas is found. Since China's border follows approximately the crest of the main range, most of these southern-facing slopes are in Nepal, Bhutan and India, but not all. In some places the border cuts across deep south-flowing valleys, and through these the dense subalpine and temperate forests are able to reach into China. Examples are the Arun and Zhangmu Kou'an valleys flowing into Nepal, the great Chumbi valley between Sikkim and Bhutan, and a very large area of the southern Himalayan faces in the Medog region of south-east Tibet.

Such areas may seem small and insignificant on the whole map of China, but in terms of biological diversity they are of enormous importance and value. There is a complete spectrum of vegetation zones, with oak-dominated broadleaf forests giving way to mixed conifer forests between 3,000 and 4,500 metres (9,800-15,000 feet). Hemlocks and spruces predominate, with a distinct fir zone appearing at the treeline. Bamboos and rhododendrons make up the bulk of the understorey, and a juniper and rhododendron scrub occurs above the treeline. Higher than this the flora becomes gradually more sparse but with a wealth of alpine genera, such as primulas, poppies, louseworts and asters, providing a riot of summer colour.

Melting glaciers and a generally moist climate, especially at the eastern end of the Himalayas, maintain moist conditions that allow for abundant vegetation growth, whilst the clear air and steep gradients provide some of the most outstanding scenery in all China.

Land profiles are steep and other geographical and climatic conditions vary sharply over small horizontal distances. Climate also varies dramatically with season. As a result, many of the faunal species of the Himalayas are altitudinal migrants. In summer, species range to high altitudes, but in winter they retreat back down into the lower valleys. Many of the birds have seasonal shifts of 1,000 to 2,000 metres (3,280-6,500 feet) in

ABOVE The Zhangmu Kou'an gorge, slicing through the Himalayas, is one of the few regions of Tibet to receive moisture from the southern side of the mountains. The result is an explosion of greenery and spectacular waterfalls.

BELOW The Slaty-backed Forktail (*Enicurus schistaceus*) patrols its territory along mountain streams, perching on prominent rocks, flicking its long tail and giving shrill calls.

altitude and some of the larger mammals, such as Takin, Blue Sheep, deer, Snow Leopard, bears, wild pigs and others, also show such migrational patterns. Small mammals that cannot make such big movements tend to hibernate through the winter, a habit seen among the flirty marmots, picas and voles. Bats are both migrants and hibernators.

Spring arrives fast in the Himalayas. One day the ground is frozen and covered in a few centimetres of snow, the next it has melted and primulas are already in flower, having developed quietly in self-warmed igloos under the snow just waiting for the sunshine to release them. A mass of other alpine flowers follows quickly. Early morning streams flow clear and pure, but by afternoon they are dirtied by the soil embedded in melting glaciers which refreeze each night.

White-collared Blackbirds emerge from the shelter of the forests to inhabit the open alpine meadows, and redstarts, chats and snowfinches add to the flutter of wings. Busy marmots clean out their burrows and frisk over the grassy slopes – one eye out for food, mates and competitors, the other always watching the skies for the frightening sight of the Golden Eagles, Lammergeiers or Himalayan Griffons that rely on their tasty flesh for their own survival.

Many species of large, colourful rhododendron bushes flower in spring and early summer, those of the lower zones coming into bloom first, followed in turn by different species in higher altitude bands. As they flower, the bees and other insects rise up the mountains with them, closely followed by nectar-feeding birds, such as the colourful Gould's Sunbird (*Aethopyga gouldiae*), the amazingly long and red-tailed Fire-tailed Sunbird (*A. ignicauda*) and the no less beautiful Green-tailed Sunbird (*A. nipalensis*), plus many insect-feeding species such as warblers and flycatchers.

Scurrying among the rhododendron roots are parties of Blood Pheasants, gaudy tragopans and spectacularly coloured Himalayan Monals. Less gaudy, but much more vocal, are the parties of laughingthrushes – several species with slightly different niches – which feed on the forest floor but always leave one or two watch-dogs in the canopy to alert the group of danger. Once warned, the whole party slips quietly away before bursting into cackling alarm calls from a safe distance. Rufous-bellied and Great Spotted Woodpeckers move into the upper flower zones, hammering out their territories, excavating breeding holes and feasting on the large wood beetle larvae that are coming to the surface to pupate before their own adult life in the summer months begins.

In the warmer weather, the forests and meadows also come to life with dazzling varieties of butterflies. The Common Swallowtail (*Papilio machaon*) is among the most conspicuous and beautiful but there is a wealth of other lovely species such as the rare spotted Apollo butterflies of the alpine meadows, Queen of Spain Fritillaries, Bath Whites, hairstreaks, skippers, tiny blues and, at lower altitudes, large Silver-washed Fritillaries and tropical swallowtails. Two species are of special interest. The gorgeous *Taeniopalpus imperialis* is truly one of the most beautiful butterflies in the world whilst the gaudy elongated *Bhutanitis lidderalidae* is one of the most valuable.

OPPOSITE PAGE The fantastic wildlife of the Zhangmu Kou'an gorge is a total contrast to the desolation of the Tibetan plateau, providing Tibet with a large proportion of its animal and plant species.

FOLLOWING PAGES Mount Everest seen from the north face base camp at 5,200 metres (17,060 feet) in southern Tibet.

Qinghai Lake

Qinghai lake, in the north-east of Qinghai province, has an area of 4,426 square kilometres (1,708 square miles), making it by far the largest semi-saline lake in China. The surface is some 3,000 metres (9,800 feet) above sea level and the lake is surrounded by grassy meadows which are densely stocked by domestic animals. The temperature is warmer than the surrounding country, averaging about 20°C (68°F) and rainfall is about 350 millimetres (14 inches) per year.

A nature reserve has been established at this inland sea to protect the thousands of breeding birds that nest on the lake every summer. Cormorants crowd onto the rock pinnacles whilst the splendid Great Black-headed Gulls and commoner, smaller Brown-headed Gulls breed in vast colonies on the lake's beaches. There are also important breeding colonies of Bar-headed Geese and pretty Pied Avocets, whose fluttery flight, smart black-on-white plumage and delicate upturned grey bills make them a birdwatcher's favourite wherever they are found.

To the east of the lake, the salty marshes and grasslands extend for many kilometres. Here the Russian explorer Nikolai Przewalski pitched his tents a century ago and observed the herds of Chiru and Goitered Gazelles that still feed on these open plains. Picas dig complex burrows whilst hamsters pack their cheek-pouches with rich grass seeds and the mole-like Zokors tunnel endlessly in search of roots and tubers. Above ground, the perky Mongolian or Henderson's Ground Jay waits for the worms to peek into the daylight, then pounces to tug them off to its hungry chicks.

BELOW Grasslands on the western shores of Qinghai lake. This area, close to the bird colonies of Bird Island, has been protected as a nature reserve.

OPPOSITE PAGE, ABOVE Common Cormorants (*Phalacrocorax carbo*) nesting on one of the large stacks in the Qinghai lake nature reserve.

BELOW LEFT The abundant Black-lipped Pica (*Ochotona curzoniae*) is a rabbit with short ears. It lives in elaborate systems of burrows and, despite popular belief, is beneficial to the grassland ecology.

BELOW RIGHT A clambering clematis (*Clematis orientalis*) grows through the herb layer in the marshy grasslands at the edge of Qinghai lake.

Kunlun Mountains

The Kunlun mountains area is cool and dry in summer and severely cold in winter, with strong winds blowing through the spring and winter months. The frost-free summer lasts only 20 to 60 days and the higher peaks have frosts all the year round. Stony and sandy deserts occur in the Qaidam basin but the rest of the Kunlun range is characterized by cold montane steppe vegetation and alpine desert supporting shrubs such as *Ajania tibetica* and *Ceratoides compacta*, with grasses such as *Stipa subsessiliflora* in diluvial fans. In the extensive Akesayqin desert area there is a vast expanse of white sand without any flowering plants at all.

Few animals can exist in this inhospitable zone. Wild Ass can live in the Qaidam basin and parts of the vegetated plateau, and can subsist on salt-tolerant plants, whilst the great shaggy Wild Yaks persist in the cold alpine desert areas. Tibetan Gazelles wander in large flocks, constantly flagging their tails, seeking the tiniest of plants that live in the stony soils, but otherwise the only mammals are a few picas and marmots, and even birdlife is rather scant.

Bare mountains form colourful backdrops to rolling barren plains and arid hills. All would remain untouched by man were it not for rich deposits of oil below the frozen ground. Now there are towns, roads and pipelines and with the influx of hardy workmen come their even more hardy rifles, so that even in this most desolate of wildernesses wildlife must forever flee.

In the centre of Qinghai the outermost feet of the Kunlun mountains come down to meet the gravel desert of the Qaidam basin, a massive graben that fills the northern half of the province.

ABOVE Reaching to about 6,000 metres (19,680 feet) the highest ramparts of the Kunlun mountains catch what little moisture there is in the air, resulting in permanent snow and glaciers in a landscape that otherwise consists of nothing but desert. They also serve to wall in the northern limits of the highest parts of the Tibetan plateau.

RIGHT The Kunlun mountains are a tortured landscape, entire mountainsides showing the scars of massive uplifting and tilting, a process that continues to this day as the Indian subcontinent pushes up against Eurasia.

ABOVE In a few valley areas of the Kunlun hardy *Artemisia* shrubs can survive, giving a patchy cover, but everywhere conditions are arduous in the extreme.

LEFT A desert landscape in the Kunlun region of central Qinghai, the valley given some life by a meandering stream fed by snowmelt from glaciers on the highest mountains some 30 kilometres (20 miles) away.

BELOW A splash of colour from a flowering *Oxytropis tatarica* shrub on bare rocky slopes just below the Kunlun pass, gateway onto the highest parts of the Tibetan plateau.

ABOVE A herd of Chiru (*Pantholops hodgsoni*) at the Aqik lake of the Arjin Mountain Nature Reserve, at over 4,000 metres (13,120 feet). Their long, rather straight horns are sought after by Tibetan herdsmen, and make a good gun tripod to help shoot more of these antelopes.

BELOW Yaks keep warm through the winter because of their dense woolly coats. Domestic herds are shorn each summer and the wool is made into tents, cloth and bindings. Their milk is made into cheese, yogurt and butter.

Two races of Wild Ass (*Equus hemionus*) occur in China. The lowland race, shown here, is found in the Gobi deserts and grasslands of Inner Mongolia. The other, the Kiang, lives in the cold deserts of the high plateau.

Focus on
South-west China

The south-west China unit covers an area of some 767,000 square kilometres (300,000 square miles) in the south-east corner of Tibet and areas of western Yunnan and Sichuan. This is biologically one of the most interesting parts of the country, having a generally moist climate and a great range of altitude and landform in some of the most rugged and fascinating scenery in the world.

Elevations range from below 1,000 metres (3,280 feet) in the valleys, which support temperate broadleaf forests, to over 6,000 metres (19,000 feet) on the highest ridges. Successive altitude zones support mixed broadleaf-conifer forests, dense stands of firs and hemlocks, juniper and rhododendron scrub, alpine meadows and scree. Slopes are everywhere very steep and there are many raging white-water rivers in deep gorges.

Three of the world's major rivers – Salween (Nujiang), Mekong (Lancangjiang) and Yangtze (Changjiang) – rise in these mountains to flow in quite separate directions. This has caused a great enriching of the local flora and fauna as climatic fluctuations of the past few thousand years have funnelled together species from a wide area of the Asian landmass.

Both flora and fauna are closely related to, and partially derived from, those of the Himalayas, particularly at higher altitudes, but there is also some mixing of elements from the Burmese, Indo-Chinese and east Chinese sub-regions, as well as a solid core of local endemics that include some of the rarest and most splendid wildlife of our planet – not least, Giant Pandas and Golden Monkeys. This is also the world distribution centre for the pheasant family, with many spectacular local examples – Golden Pheasant (*Chrysolophus pictus*), Blue and White Eared Pheasants (*Crossoptilon auritum* and *C. crossoptilon*), Monals (*Lophophorus sclateri* and *L. ihuysii*) and Temminck's Tragopan (*Tragopan temminckii*), to name a few.

The alpine flora is the most diverse in the world, as are the high-altitude forests. For example, the Wolong nature reserve of the Qionglai mountains alone boasts a list of some 4,000 vascular plants. Within a few kilometres you can encounter many different species of such familiar domesticated fruits as strawberries, blackcurrants, gooseberries, cherries, plums, apples, pears, brambles; not just one species of each but many varieties of each type, constituting a remarkable treasure of usable genetic materials for future cultivar improvements. It is easy to see the importance of preserving the biodiversity of this unique region.

The reason for the amazing abundance and high endemicity of this unit lies in its geological history, physical geography and climate. These mountain ranges mark the eastern fanning of the Tibetan plateau, which has been raised only recently in the geological timescale as a result of tectonic pressures. This gradual rise caused the species that lived around the ancient Tethys Sea to become extinct or migrate outwards, maintaining their altitudinal range as their original home became raised to 4,000 metres (13,000 feet). Some of these ancient relict species have survived here.

The raised plateau acquired new climatic features, including extreme high pressure for most of the summer months and extreme aridity. Although the south-west monsoon brings plenty of mist towards the plateau, this moisture is unable to rise high enough to come onto the top of the plateau. Instead, the moisture-laden clouds build up in the valleys and on the mountain ranges to the east, giving very cloudy wet summers. There is a saying in Sichuan that 'dogs bark at the sun' so rarely is the sun to be seen in the summer months. The wet forests grow dense thickets of bamboos as an understorey, which provides good habitat for the Giant and Red Pandas. In the moist summers leeches emerge to patrol the forest trails and suck the blood of any passing mammals.

With the climate fluctuation of the Pleistocene period during the last three million years, much of the ancient flora of the Palaearctic region was decimated by periodic freezing or desiccation, though in this part of China it was able to survive intact. Even in the driest periods of the Pleistocene, whatever moisture there was would become trapped in these last valleys, unable to rise to the plateau. There were no major droughts and any temperature fluctuations could easily be accommodated by minor adjustments in altitude which involved only tiny horizontal movements as a result of the steep land profile.

Another factor contributing to high levels of endemism is the dissected nature of the landscape. Valleys of similar climate may be separated by high ridges which totally isolate one from another. The downward migration route for many species of temperate flora is equally blocked, as the subtropical ecosystems of lower altitudes are as inhospitable to them as the high, glaciated ridges above. Each temperate valley behaves like a habitat island, evolving its own endemic forms, subspecies and eventually species. However, when vegetation zones have moved up the mountains in warm periods and lower down in cold, periodic linkages have occurred between otherwise isolated valleys, building up the overall species richness. Thus, many species could spread around the upper valleys of the Yangtze (Changjiang) river and eastwards along the crescent of mountains to the Qinling mountains of Shaanxi, and some as far as the forested peaks of Shennongjia in Hubei province.

The south-west unit is now very heavily populated by humans and the local farmers have developed agricultural systems that make use of all land types. Valley bottoms are irrigated and hills are terraced, whilst the wilder slopes provide grazing for domestic yaks and cattle, firewood for the cold winters and hunting areas to supplement the kitchen. More recently, heavy logging of the valuable timber has caused widespread loss of wildlife habitat. Orchards or scrub have replaced the once dense forests, the endemic fauna is endangered and the urgency for solid conservation efforts is extreme, though several reserves are already making a notable contribution.

Wolong, in the Qionglai mountains, is the largest nature reserve in the unit. It includes an enormous altitudinal spectrum from the subtropical broadleaf forests at 1,100 metres (3,600 feet) right through the temperate broadleaf, mixed conifer-broadleaf and coniferous subalpine zones to the alpine meadows and pastures above the treeline at 3,300 metres (10,800 feet), and then the permanent ice and rock of the beautiful peak of Siguniang ('Four Maidens') at 6,250 metres (20,505 feet).

The main road through the reserve climbs up the Pitiao valley and then winds up to the pass of Balangshan at 4,487 metres (14,721 feet) before passing over into Baoxing county, formerly the kingdom of Mupin, and out of the reserve. The Balangshan pass offers the easiest access to the alpine zone of Wolong and has a lush alpine flora with slipper orchids (*Cypripedium*), dazzling blue *Corydalis*, a range of purple louseworts (*Pedicularis*) and many species of wild poppies (*Mecanopsis*). The meadows of Balangshan were described with great enthusiasm by the famous British botanist Ernest White, who did so much of the early botanic exploration of west Sichuan before his premature death in a car crash early this century. He left behind a series of photographs of Balangshan and it is with some surprise but much consolation to see that the site looks just the same today as it did 90 years ago.

Giant Pandas live in Wolong at almost all altitudes up to the treeline. They show seasonal shifts in ranging behaviour, using

OPPOSITE PAGE Hailuogou glacier streams down the slopes of Gonggashan, Sichuan's highest mountain and a massive easterly outrider of the Tibetan plateau.

ABOVE Mountains of northern Sichuan, near the town of Songpan. In areas such as this the government has agreed to ban logging to improve protection of the Giant Panda.

BELOW LEFT Early morning fog shrouds primary coniferous forest on the slopes of Gonggashan.

BELOW RIGHT There are several reserves around Gonggashan; the scene shown here is in Hailuogou Glacier Park, which protects a spectacular mix of forest, alpine slopes and high mountains, with the Hailuogou glacier carving through it all.

ABOVE LEFT The Dadu river rumbles downwards through a spectacular gorge just south of Gonggashan. The Dadu rises in the mountains of western Sichuan and travels virtually its entire course down gorges, much of the time as white water, before finally joining the Min at Leshan.

ABOVE RIGHT A flower of the Yunnan Pine (*Pinus yunnanensis*). The spores are dispersed by wind.

LEFT Spring is the best time to visit Wolong Nature Reserve. The weather is clear, azaleas (*Rhododendron augustinii*) are in full bloom and the sunlight still reaches the forest floor.

OPPOSITE PAGE, LEFT The Golden Pheasant (*Chrysolophus pictus*) is regarded as one of the most beautiful birds in the world, but this does not stop the males from risking their looks in a territorial scrap when challenged.

OPPOSITE PAGE, RIGHT The Silver-eared Mesia (*Leiothrix argentauris*) is a colourful and noisy flocking bird of scrub and forest undergrowth. It is an inquisitive species and easily attracted by imitation owl calls or 'pishing'.

the lower deciduous forests in winter and spring, feeding on umbrella bamboo (*Bashania*) leaves in winter and the new shoots in the spring, but moving up into the high conifer forests for most of the summer and autumn to feed on the much smaller arrow bamboo (*Fargesia*).

Two other reserves in Sichuan also hold Giant Pandas but are worth visiting for other reasons. Jiuzhaigou is less rich biologically than Wolong but more scenic. The reserve lies in a steep area of the Minshan mountains where sharply inclined faults have resulted in landslips, creating amazingly beautiful lakes, waterfalls and clear rivers. The peaks are sharp and conical, largely bare limestone and quartz rocks, whilst the upper slopes are heavily covered with virgin conifer forests of stately hemlocks, spruce, fir and white pines.

Emeishan, in southern Sichuan, is an isolated mountain of upthrust limestone rising to some 3,099 metres (10,167 feet). Some of the faces of the peak rise almost sheer for over 3,000 metres (10,000 feet). The mountain is home to over 70 monasteries scattered along a narrow trail that winds up through the forest to an alpine meadow of dwarf bamboo thickets. The trip to the summit usually takes more than a day, necessitating staying the night in one of the monasteries. On a clear day, when clouds of mist float in the valleys, an aureole like a floating golden ball surrounded by a rainbow may sometimes be seen. This phenomenon is known as *Foguang* or 'Glory of Buddha'.

The south-west China unit is the focal point of Chinese bird endemism. Many species with restricted ranges are found in Sichuan and adjacent provinces. Among these, the Sichuan Partridge (*Arborophila rufipectus*) is known only from south Sichuan. The bird resembles the Hill Partridge but has a white forehead and russet breast. The White-backed, or Kessler's, and Chinese Thrushes (*Turdus kessleri* and *T. mupinensis*) are both confined to upper forests. The latter looks much like a Song Thrush but the White-backed resembles a Blackbird with rufous mantle, whitish back and rufous underparts. Four laughingthrushes are restricted to this unit. The Barred Laughingthrush (*Garrulax lunulatus*), decorated with bold black

crescents, is distributed throughout the same mountain ranges as the Giant Panda. The closely related White-speckled or Biet's Laughingthrush (*G. bieti*) is found just to the south, in south-west Sichuan and north-west Yunnan. The much larger Giant Laughingthrush (*G. maximus*), identified by its black and white spots, is found from south Gansu, through west Sichuan to north-west Yunnan, south-east Tibet (and just across the border of north-east India); the drabber Elliot's Laughingthrush (*G. elliotii*) is very common at moderate altitudes throughout the unit.

Emeishan (sometimes spelled Omeishan) is the centre for its own cluster of very restricted endemics. Here we find the colourful Emeishan Liocichla (*Liocichla omeiensis*), the flock-living Golden-fronted Fulvetta (*Alcippe variegaticeps*) and the little Grey-hooded Parrotbill (*Paradoxornis zappeyi*). The streaky-breasted Chinese Fulvetta (*A. striaticollis*) lives with another endemic, the White-necklaced or Sooty Tit (*Aegithalos fuliginosus*), in the prickly oak zone above the Giant Pandas, whilst the Rusty-throated or Przewalski's Parrotbill (*Paradoxornis przewalskii*) is confined to the northern parts of the unit, along with the White-browed Tit (*Parus superciliosus*). The pretty Rusty-bellied or David's Tit (*P. davidi*), the Spectacled Parrotbill (*Paradoxornis conspicillatus*), with its slim white eye-ring, and the Three-toed Parrotbill (*P. paradoxus*) also occur in the same habitat as the Giant Panda. The last is a large parrotbill with a floppy grey crest and broad white eye-ring.

The pale-bellied Yunnan Nuthatch (*Sitta yunnanensis*) is found in Yunnan and south Sichuan whilst the Pink-rumped Rosefinch (*Carpodacus eos*) is found mostly at the northern end of the unit in Minshan and the south-east corner of Qinghai. The Three-banded Rosefinch (*C. trifasciatus*) is another bird which occurs in the same range as the Giant Panda, emphasising yet again what a good 'flagship' species the panda makes for conservation. This species differs from most of the other rosefinches by having a white belly. One curious endemic bird of the unit is the Sichuan Jay (*Perisoreus internigrans*) which is all grey with a greenish bill. This jay lives in the upper conifer forests feeding in trees and on the ground but including conifer seeds in its mixed diet.

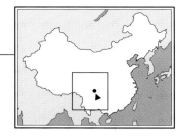

Emeishan

The sacred mountain of Emeishan is situated quite close to the famous eighth-century Giant Buddha of Leshan, a huge figure carved into the cliffside overlooking the Min river. Mixed deciduous and coniferous forests occur in this reserve from 1,100 metres (3,600 feet) up to 1,900 metres (6,200 feet), above which the forest become dense and dark, pure coniferous stands of firs and hemlocks. These forests are alive with Tibetan Macaques, woolly brown scamps that have learned to molest tourists and pilgrims for food. There are also rare sightings of tracks of Giant Pandas on the back of the mountain.

Apart from the scenic beauty and religious significance of Emeishan, the reserve has special importance for Chinese fauna as it is the type locality of several very localized endemic species. Perhaps the most spectacular of these is the Emeishan Liocichla, a small, brightly coloured kind of laughingthrush.

The mountain has been well explored by, among others, the famous botanists Ernst Faber and E.H. White. Faber's first collection of plants from here contained no fewer than 70 new forms. The evergreen tree *Nothophoebe omeiensis*, the white-flowering herb *Sibbaldia omeiensis* and the rose *Rosa omeiensis* are splendid examples. But Emeishan also has some valuable specimens of wider-ranging rarities such as Dove-trees, the Spur-leaf and sweetgums.

The 3,099–metre (10,167–foot) summit of Emeishan – one of China's four Buddhist sacred mountains – is a spectacular sight, falling away to the east in a cliff that in many places is sheer for nearly 1,000 metres (3,300 feet).

ABOVE The Emeishan Liocichla (*Liocichla omeiensis*) is one of China's rarest birds with a very limited distribution centred on its namesake mountain. Liocichlas are related to laughingthrushes but are identified by a very square-cut tail. The Emeishan species is prettily touched up with red edges.

RIGHT Tibetan Macaques (*Macaca tibethana*) are quite common on Emeishan. They live in troops of 20–50 animals and pester visitors for food. They are shorter-tailed, longer-haired and a darker brown than the Rhesus Macaques which occur further east and in the Himalayas. BELOW RIGHT An infant shows off away from its mother. At this age it is vulnerable but the mother and others in the group will keep a close watch. Any predator threatening the baby would find itself attacked by the group's fiercest members.

BELOW LEFT The moist conditions on Emeishan provide an ideal place for a variety of forest ferns. Three different species can be seen growing together in this photograph.

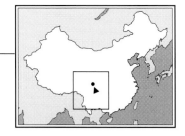

Wolong – Home of the Giant Panda

Wolong Nature Reserve, with an area of 50,000 hectares (123,450 acres), is the largest of some 27 nature reserves established to protect the habitat of the Giant Panda, and was established in 1963. It can be reached in three hours by car from the Sichuan capital city of Chengdu.

The headquarters of the reserve are at Shawan in the Pitiao valley, but a few kilometres downstream at Hetaoping is a research laboratory and panda breeding centre established jointly by the Ministry of Forestry and the World Wide Fund for Nature (WWF). Here a number of captive pandas have been kept for breeding, with moderate success. It is hoped eventually to try releasing captive-born pandas back into the wild where the natural population is at reduced density as a result of the mass flowering and subsequent die-off of the bamboo species on which these animals depend, which occurred in the mid-1970s and again in the early 1980s.

Although Wolong has many management problems due to the presence of some 4,000 Qiang minority farmers, and some major hydro-power developments in the Gengda valley, it is none the less an absolute treasurehouse of China's endemic biodiversity.

Above Shawan, at an altitude of 2,800 metres (9,200 feet), is the research camp of Wuyipeng where much of the pioneer field study of Giant Pandas was undertaken in the early 1980s. The original tents have been augmented with several metal huts and electricity has been supplied to make life in the camp during the winter snows more bearable. The station is still used for continuing field research on the pandas, bamboos, pheasants and other wildlife.

The Pitiao river of Sichuan's famous Wolong Nature Reserve runs through the temperate broadleaf zone. The reserve spans a huge altitudinal range from permanent glaciers to subtropical evergreen forest. The Giant Pandas for which it was established live mostly in the subalpine conifer zone, where the bamboo understorey is thickest. Wolong also boasts the greatest diversity of plants and birds of any non-tropical reserve.

ABOVE The Giant Panda (*Ailuropoda melanoleuca*) has an opposable thumb-like lump on its hand to help it hold the stems of bamboo. After mass flowerings and subsequent die-back of the bamboos in 1973 and 1984 only about 75 pandas survived in Wolong. Now the bamboo is recovering and the panda population is on the increase.

RIGHT Giant Pandas often climb trees to get some rare sunshine, to rest in peace or sometimes to escape from man or predators below. They scramble down awkwardly, less arboreally agile than true bears.

The alpine zone of Balangshan, in the Wolong reserve, is splashed with colourful flowers. ABOVE LEFT The Lampshade Poppywort (*Mecanopsis integrifolia*) may stand up to a metre (3 feet) tall and tower above the other herbs.
BELOW LEFT The purple *Primula nivalis* flowers early in spring, soon after the covering snow has retreated.

ABOVE RIGHT Marsh marigolds (*Caltha*) crouch close to the damp ground.

BELOW RIGHT The fluted flowers of *Corydalis* aff *flexuosa* attract small nectar-feeding butterflies.

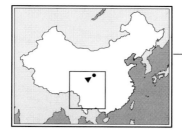

Jiuzhaigou

Jiuzhaigou, in the Min mountains in the far north of Sichuan, has become a famous destination for tourists, both international and domestic, not so much because of its wildlife, although indeed it has Giant Pandas, bears, Golden Monkeys, leopards and splendid pheasants, but because of its exceptionally beautiful scenery.

Several large elongated lakes at the south of the reserve stretch languidly into the distance, feeding clear mountain streams that cascade northwards to form smaller secondary lakes and series after series of gorgeous waterfalls. Some of the small lakes are exceptionally clear and still, and acquire bluish, greenish and even purplish tints from the lime and minerals in the rocks. Dead trees and roots form weird sub-aquatic gardens whilst the surrounding meadows of *Ligularia przewalskii* add bands of bright yellow to an already fairy-tale scene.

Rainbows form around the base of the great Pearl Shoals falls, where visitors jostle for the prized vantage points for those precious photos which will form the only mementos of a rare break from the factory floor or school classroom.

Decoratively dressed Tibetan villagers lead their saddle horses up the Baihe valley to their hamlets, where water-driven prayer-wheels deliver their petitions nonstop under the fluttery prayer flags. Tiny fields, often only a few metres square, cling to any part of the hills with a shallow enough gradient to hold a little soil and a crop of buckwheat.

The Jiuzhaigou river drains north past the entrance of the reserve, swings east and then joins the Baishuijiang river, sweeping back to the south as the Jialing river to mix with the Yangtze and start the long journey down through the gorges to Shanghai some 2,000 kilometres (1,200 miles) to the east.

The Shuzheng falls are among an extensive series of waterfalls that can be seen in the beautiful valleys of Jiuzhaigou. Many of the rivers that feed these falls are characterized by being wide and very shallow, creating hundreds of small islands crowded with shrubs and trees (as can be seen here) and resulting in natural water gardens.

ABOVE A classic Jiuzhaigou view: a small but stunningly blue lake, just one of many strung along the reserve's two valleys, surrounded by the rich greens of mountains clothed in mixed broadleaf-coniferous forest. The splashes of purple are wild azaleas in full flower.

BELOW Nuorilang falls, yet another jewel in the beauty of Jiuzhaigou, sit at the meeting point of the valleys. From here, the reserve continues northwards as a single valley, filled with lake after lake, to the entrance where the Jiuzhaigou river swings eastwards.

ABOVE A small section of the base of the magnificent Pearl Shoals falls, Jiuzhaigou's star attraction. The water cascades down as a multitude of interconnecting ribbons broken up by trees and shrubs.

BELOW Mirror lake sits on a shelf in Jiuzhaigou's mountains, fed by Pearl Shoals falls and drained by Nuorilang falls, surrounded by mixed forest and snow-capped mountain peaks.

ABOVE RIGHT The pretty blue butterfly *Lycaeides argyrognomon* settles delicately on a spray of leaves.

ABOVE LEFT Maples in Jiuzhaigou turn yellow in the early autumn. By late September they will be a startling scarlet to turn on more magic in this outstanding landscape.

LEFT Beautiful rotund Lady's Slipper Orchids (*Cypripedium franchetii*) sport pregnant globes among the early summer alpines.

BELOW LEFT *Lycopodium* and *Sphagnum* mosses cover the damp forest floor in Jiuzhaigou, making it difficult for tree seedlings to take root in the soil.

BELOW CENTRE The whiskery flowered *Paris polyphylla* herb is a valuable plant used in China to make the famous Yunnan white powder medicine.

BELOW RIGHT Huge cabbage lichens grow on the tree branches of the montane forests and form much of the diet of the Golden Monkeys.

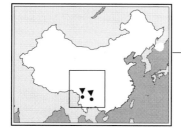

North-west Yunnan

In north-west Yunnan, where the three valleys – Salween, Mekong and Yangtze – start to open out and diverge, there are several small lakes. Here Black-necked Cranes migrate each winter from their swampy breeding grounds on the cold Tibetan plateau. Swans, geese and other waterbirds winter on the water surface, though the cranes prefer the moist land around the valley heads.

To the east and centre of north Yunnan the countryside becomes more opened up by agriculture. There is not much forest left. Junipers grow on limestone ridges where the soil is too poor to cultivate and extensive areas are planted with pines – Yunnan Pine in the north and Simao Pine in the south. Here and there natural forests still survive on slopes such as the Western hills south of Kunming city. In such relict forests you can still see many interesting and rare birds, such as the Giant Nuthatch.

Perhaps the strangest place in north Yunnan is the so-called 'stone forest', or *shilin*, some three hours east of Kunming, where an extensive area of karst limestone has been weathered into thousands of narrow vertical pillars of dolomite. It is a curious geological formation and a few specialist plants enjoy the safety of the more inaccessible pinnacles. Tourists flow into the western end of the reserve to be photographed with colourful Sani minority girls, but otherwise this is not an area teeming with wildlife.

The famous 'stone forest' of north Yunnan is an area of karst limestone. Hard minerals protect the softer limestone directly beneath whilst the rest becomes eroded by rain, leaving thousands of sharp pinnacles of rock which grow taller as more surrounding earth is worn away each year.

ABOVE The yellow-flowering *Berberis* is a common shrub of open country. It makes a hardy, decorative garden plant but is used locally for medicine.

RIGHT The jagged peaks of 5,596-metre (18,360-foot) Yulong Xueshan ('Jade Dragon Snow Mountain') tower over the nearby town of Lijiang in north-western Yunnan. This mountain, which marks the most southerly permanent snow and ice in China, forms the eastern wall of the mighty Tiger Leaping Gorge through which the upper Yangtze is funnelled, and has defeated all attempts to climb it.

BELOW A new mountain stream starts its life as it flows from a spring on the slopes of Mount Yulong Xueshan.

LEFT AND ABOVE A *Cotoneaster* bush overhangs a river in north-west Yunnan. The genus is immensely popular in the West as a cultivated plant and its bright berries enliven many a park and garden during the autumn.

BELOW LEFT The Red-headed Tit (*Aegithalos concinnus*) is a common and agile bird of forest and woodland, living in busy flocks that work their way through the canopy catching insects and eating small seeds.

BELOW RIGHT A delicate azalea, *Rhododendron* aff *polylepis*.

BOTTOM The rich creamy flowers of *Rhododendron wightii*.

The north-western Yunnan region, like many other areas in China, is rich in attractive rhododendrons. The smaller, rather fragile-looking species known as azaleas lose their leaves in winter, unlike the larger evergreen species, but are all included within the genus. These beautiful plants are much prized in cultivation for their massed bunches of flowers that are displayed in spring and early summer.

ABOVE LEFT AND RIGHT Pink and scarlet varieties of *Rhododendron arboreum*.

BELOW LEFT The charming azalea *R.* aff *lutescens*.
BELOW RIGHT A dwarf rhododendron in the Western hills south of Kunming.

FOCUS ON
CENTRAL CHINA

This is a large region comprising most of China south of the Yangtze (Changjiang) river from the east coast as far west as central Sichuan and Yunnan, but excluding the southernmost tropical fringe.

Much of the land is mountainous but some of the wider valleys have alluvial plains. The 260,000-square-kilometre (100,000-square-mile) Sichuan basin is a large, flattish, alluvial lake-bed with a series of parallel limestone ridges. Despite the generally rugged terrain, the unit contains a huge human population of over 300 million people, more than 100 million of whom live in the fertile Sichuan basin.

To the south of the Sichuan basin is the Guizhou plateau, with altitudes generally between 1,000 and 2,000 metres (3,280–6,560 feet). This region is very complex and has been described as 'without three feet of continual level land, without three successive days of fine weather'. What gives the area such an irregular surface is the extensive karst limestone which has eroded into a landscape of spectacular pillars and honeycombs. Where rivers meander through such geological formations one finds amazing scenery of the kind so fancifully depicted in many Chinese paintings.

Karst limestone forest is characterized by having high levels of endemism in the flora, snail fauna and some other groups such as birds and monkeys. Two rare monkeys – White-headed and Black-headed Leaf Monkeys (*Trachypithecus poliocephalus* and *T. francoisi*) – are found in this area and protected in special reserves.

The eastern parts of the unit are also mountainous with the three ranges Nanling, Xuefeng and Wuyi forming a semi-circle, enclosing the flatter lands to the north which grade down through rolling hills into the Yangtze valley. The famous Yangtze river, generally known in China as Changjiang or 'long river', is a major feature of the region. This is the third longest river in the world after the Amazon and Nile. From its source in the high mountains of Qinghai province to its mouth by the great town of Shanghai, it runs for 6,300 kilometres (3,915 miles) and, together with its 700 tributaries, drains much of six provinces with a catchment area of 1.8 million square kilometres (well over ½ million square miles). Each year the river discharges almost one billion cubic metres of water into the South China Sea, carrying with it tons of silt far out to sea.

At over 6,600 metres (21,600 feet) in the Tanggula mountains on the Qinghai-Tibet border, the source of the Yangtze is a barren alpine landscape of glaciers and snowfields pitted with moraines, swept by fierce winds and devoid of vegetation. For the winter months all moisture is locked in ice and even in summer the frost is permanent a few centimetres below the surface. Only in the heat of the day does the snow and ice of the surface melt and the first trickles of the mighty river run ever faster into the collecting streams of the Tuotuo river.

Below 5,000 metres (16,000 feet) the first hardy alpine flowers appear – tussocks of harsh grasses, stunted potentillas, colourful liverworts. Wild Yaks, Mongolian Gazelles, Tibetan Antelope and Blue Sheep feed on the sparse vegetation, keeping a sharp eye out for the cunning Lynx and beautiful Snow Leopards that prowl the mountains. As the streams descend and congregate the vegetation becomes lusher and marshy. Geese and White-lipped Deer gather around small lakes and black yak-wool tents and wisps of smoke mark the camps of the Yushu herdsmen, guarding and milking their yak herds in the summer pastures.

Side branches join the main river, whose name is now Jinsha or 'golden sands', which flows fast to the south defining the natural boundary between Tibet and Sichuan province. Here it parallels the great Lancangjiang and Nujiang rivers before the town of Shigu when, with a sudden reversal, the swelling waters turn back to the north, and wind a zigzag route through the Hengduan mountains of northern Yunnan, continuing eastwards through southern Sichuan province to be joined by the northern tributaries – the Yalong, Min and Jialing.

By the time the river reaches the city of Chongqing it is a huge, muddy, swirling current littered with cruising logs and the debris of the 120 million farmers of the Sichuan basin. Chongqing lies on the hill that holds the Yangtze and Jialing rivers apart for the last few kilometres before the combined waters crash on down to the great gorges of the river's middle reaches. Chongqing is China's most populated city, with a population of over 15 million, and also holds the distinction of having the worst acid rain in east Asia. It is an industrial centre where sulphur-rich clouds hang in the sheltered valleys.

The three gorges of the middle Yangtze are a famous spectacle and the source of a large sight-seeing industry, with tourist boats plying between Chongqing and Yichang. In midsummer, when the Sichuan rivers are in full spate, the water level in the gorges rises by more than 100 metres (330 feet) and boat passage is very dangerous. The gorges fill with mist and spray to give a mysterious air to the towering cliffs and pinnacles.

OPPOSITE PAGE Mixed forest on the middle slopes of Mount Huangshan, one of China's most famous mountains, located in southern Anhui. With the advance of autumn the fiery red leaves of the *Sorbus* trees make a stunning contrast to the sombre green of the pines.

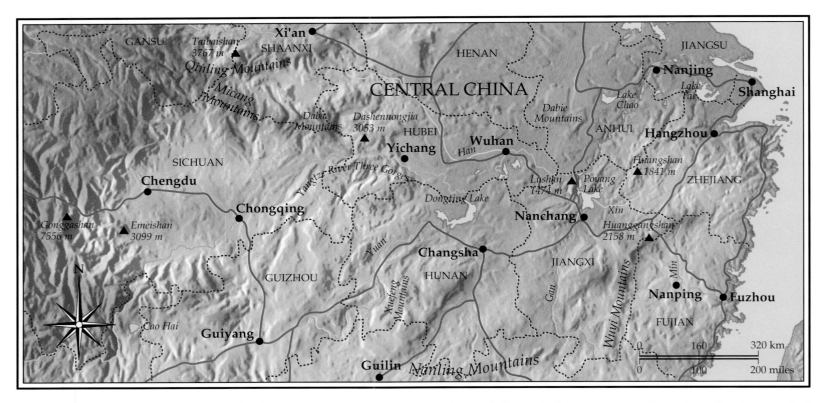

There are also some very beautiful tributary gorges, perhaps the loveliest of which is the Daning river above the small town of Wushan. Here the scenery and vegetation remain undisturbed. Peaks and pillars tower out of the clear water and the boat trip passes through gorges and shimmering rapids. At Dicui Xia there are wild ducks and, on a lucky day, the famous Golden Monkeys can be seen scampering over the rockcliffs above.

Below the gorges the river runs more smoothly to the coast, but in its lower reaches it passes many great lakes such as Poyang and Dongting. In former times there were over 50 smaller lakes connected directly to the Yangtze but, today, most have been impounded and turned into fish farms or reclaimed for agriculture. It is in the lower reaches of the river that the Yangtze dolphins, sturgeon, paddlefish and alligators live, but all are now highly endangered. The population of Chinese River Dolphins (*Lipotes vexillifer*) numbers fewer than 100 individuals. They are threatened by the barrages and fishing-nets which block their passage, as well as by the increasing levels of pollution, decrease of prey fish and direct damage from ships' propellers. A project to save the population by captive breeding has so far failed, with only one solitary animal successfully held in captivity.

Poyang and Dongting are the largest lakes of the lower Yangtze river system. Poyang, for instance, is fed by several rivers that flow from all corners of Jiangxi province. At its north end the lake is connected by a narrow gorge to the Yangtze river under the shelter of the sacred and well-forested Mount Lushan. The lake changes in size and depth with the season. In summer it fills to its maximum and when the Yangtze river is in full flood the water flows backwards from river to lake. Vast areas of agricultural land go under water, fish mix freely between lakes and rivers, and what wildlife there is must swim or make its way onto higher ground. The lake may achieve a surface area of 3,583 square kilometres (1,380 square miles) but the deepest point is only 30 metres (100 feet) and much of it is far shallower.

As autumn approaches, the lake levels drop and more and more land is revealed for cultivation. The lowest and last lands to re-emerge are left for grass and reeds. The reeds are cut for thatch and animal feed by the villagers and the grass is grazed by water buffalo, resting after a frantic season of preparing rice-fields.

Around the main lakes are a number of smaller impounded lakes that are managed as fish farms. In summer, these are all connected and the fish lakes are restocked from the main lakes and rivers. The trapped fish stocks are allowed to grow until the end of the autumn, before the farmed lakes are drained off and the fish caught and taken to market. Such is the seasonal pattern of land-use that attracts one of the world's most spectacular congregations of wintering waterfowl. Great flocks of swans, ducks and geese form floating rafts on the main lakes but spread out to feed each day on the newly exposed muddy land, which is in a period of rapid plant growth after its months of submersion.

Few mammals can withstand the frightening water regime of seasonal flooding around the Yangtze lakes but the little Chinese Water Deer (*Hydropotes inermis*) is one species that can cope. It sneaks onto the grassy mudflats in the evening to feast on the new growth, hiding up in tall reeds by daytime. An animal totally at home is the Otter (*Lutra lutra*). Formerly it was very common, feeding on the fish in the lakes and rivers and nesting in the banks of quiet upper streams. Today, the animal is rather rare due to trapping for its valuable fur and because it is regarded as a pest by the fish farmers whose stocks it tends to treat as its own.

Poyang lake is a wonderful bird refuge today, but so dynamic is the river and lake system that nothing is certain for the future. Each year the load of pollution flowing into the lake increases. Recent finds of an oil-field on the lake's eastern side pose further threat. The great increase in domestic duck farms brings the risk of disease spreading to wild waterfowl, but perhaps the biggest danger comes from the colossal plans for hydrological works to control the floods of the Yangtze river. One giant project already under way is the construction of the Three Gorges Dam upstream on the Yangtze. This dam will reduce the flood flow down the Yangtze and hence change the reverse-flow regime which floods Poyang each year. Another plan involves building a barrage across Poyang itself, to hold back its waters until the Yangtze river is at a safe level for the lake to be let out.

Any future changes in water level and the timing of water lowering will have huge effects on the attractiveness of Poyang for wintering waterfowl. It is even difficult to predict whether

earlier or later water lowerings will be a benefit or a disaster for the different species involved. All we know is that the status quo is fine for the birds and any major change should be viewed as a threat to the long-term security of Poyang Lake Nature Reserve and the waterfowl species that depend on it.

Below the lakes, the Yangtze winds through the 'nine intestine-like bends' before fanning out into a broad delta of agricultural lands, lakes, islets and reedbeds. Here it creates habitat for yet another Chinese rarity. In the salt marshes of Yancheng is one of the nesting areas of Saunders's Gull (*Larus saundersi*), a delicate species of black-headed gull which winters along the south coast as far as Hong Kong. The nesting site was not discovered until 1984, and only a few hundred of these birds survive there.

The vegetation of central China is subtropical in the lowlands and temperate on the hills and mountains. The natural lowland formation was a dense hardleaf, broadleaf forest composed of evergreen oaks (*Cyclobalanopsis*) and chestnuts (*Castanopsis*). The understorey is rich and composed of many trees and bushes of the laurel family, bamboos and even tropical genera. At moderate altitudes this formation becomes mixed with more deciduous trees such as birch (*Betula*) and oaks (*Lithocarpus*), with Chinese Pine (*Pinus massoniana*) and Chinese Fir (*Cunninghamia lanceolata*). The highest mountains have forests of hemlocks and juniper with exposed peaks almost bare of trees. Most of the lower formations have been cleared for agriculture and barren lands are now being extensively reforested with conifer plantations.

The reserve of Wuyishan spans the widest range of forest types in south-east China. In its highest parts are edaphic grasslands dotted with stunted juniper trees and alpine shrubs. A cluster of giant granite boulders marks the summit, from which

In Wuyishan Scenic Area, Nine-bend Creek twists its way among the hills of northern Fujian. This area is renowned mainly for its scenic beauty, but a short distance west lies Wuyishan Nature Reserve, a region of natural forest that is a treasure trove of biological diversity.

the visitor can gaze into the distance in every direction. Sheltered parts of the summit zone support willow thickets in the beds where small streams start to form. Below 1,700 metres (5,500 feet) the dark conifer forests start and extend down to 1,000 metres (3,280 feet). Important species include Chinese Cedar (*Cryptomeria fortunei*) and Chinese Yew (*Taxus chinensis*). Below this is a zone of intermixing conifer and broadleaf forests. Often these are indeed intermixed but, for the most part, conifer forests of *Pinus massoniana* and *Cunninghamia lanceolata* grow on the drier west-facing slopes whilst broadleaf oak forests grow on the cooler, moister east-facing slopes.

Below 1,200 metres (4,000 feet) all forests are subtropical broadleaf dominated by *Lithocarpus* species. Mixed within the broadleaf forests are tall feathery bamboos known as *mao* or brush bamboo. This species is highly valuable for many products from chopsticks to paper. Unlike other bamboos, which grow in dense stands or clumps, the *mao* bamboo spreads under the ground to emerge as single spaced stems. The foresters and villagers manage the forest by cutting other species around each emerging shoot. After many years of such management only the bamboo survives and, indeed, large areas of lowland forests in the Wuyishan Nature Reserve have been thus converted into monocultures. In some areas this is threatening the linkages between different blocks of natural mixed forest and the altitudinal migration patterns of the resident fauna of the reserve.

The fauna of the region is a blending of tropical Indo-Malayan elements such as tiger, leopard, sambar and civets, with more

ABOVE The little Chinese Water Deer (*Hydropotes inermis*) is endemic to the lower Yangtze valley. It can swim to higher ground when the river floods. LEFT Sika Deer (*Cervus nippon*) are rare in the wild but are kept in many farms, where they are reared for their valuable velvety young antlers that are cut and sold for medicine.

northerly Palaearctic species such as bears, foxes and badgers. The unit has clearly experienced times of relative isolation that have given rise to a large number of endemic species and subspecies such as the Black-fronted Muntjac (*Muntiacus crinifrons*), which occupies the mountains, and the small Reeve's Muntjac (*M. reevesii*) of the lowlands. Cabot's Tragopan (*Tragopan caboti*), with its naked yellow throat pouch, is a spectacular endemic bird, and in the Yangtze basin such endemics as Chinese Water Deer (*Hydropotes inermis*), Giant Salamanders (*Audrias davidianus*) and Yangtze Alligators (*Alligator sinensis*) are found. Many of the region's endemic mammals and birds were discovered at the localities of Guadun and Dazhulan in Wuyishan. This was a favoured collecting area of Père Armand David, and produced the type specimens of many species and subspecies such as the tiny Short-tailed Parrotbill (*Paradoxornis davidianus*), Silver Pheasant (*Lophura nycthemera*), White-necklaced Hill Partridge (*Arborophila torqueda*) and the Red-tailed Laughingthrush (*Garrulax milnei*).

The central China unit is also of great importance for migratory birds. These are the first extensive evergreen forests for northern passerines on their autumn passage south. Many stay for the whole winter but, even to those that fly on to the tropical regions, this is a vital staging area. On bleak highland lakes such as Cao Hai in Guizhou province, wintering Black-necked Cranes find protection, whilst the large lakes in the Yangtze valley, such as Poyang and Dongting, are wintering grounds for vast aggregations of waterfowl – swans, geese, ducks and the rare Siberian Crane (*Grus leucogeranus*).

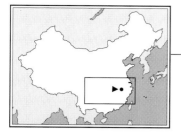

Huangshan

Situated in the most southerly corner of Anhui province, this must rate as China's most famous scenic mountain, immortalized and revered by poets and artists for centuries. The quintessential Chinese portrayals of aged and wind-gnarled pines clinging to impossibly precipitous rocky outcrops, a sea of cloud swirling far below, and jagged peaks penetrating the white veil in the background are virtually all inspired by Huangshan's landscape. In 1990, Huangshan's outstanding beauty received official international recognition, when UNESCO gave it status as a World Heritage Site.

Reaching an altitude of 1,841 metres (6,040 feet), Huangshan is by far the highest mountain in this region. The weird rock formations are composed of granite boulders, columns and pinnacles – the mountain is said to have 72 peaks, 30 of them over 1,500 metres (4,900 feet) high. The strangest of the oddly shaped pines and rocks have been given poetic names, such as the 'Rock that Flew from Afar', 'Jade Screen Peak', 'Welcoming the Guest Pine' and the 'Sleeping Dragon Pine'.

The mountain is very heavily touristed, though the vast majority of visitors head straight for the famous views around the highest peaks, getting there via cable car and leaving much of the rest undisturbed. Almost the whole of Huangshan is densely forested, and 117 square kilometres (45 square miles) of it have been designated a nature reserve. At the lower levels the forest consists of evergreen oaks and chestnuts, though by the halfway point it has become mixed coniferous-deciduous broadleaf, which in autumn comes alive with fiery splashes of red as mountain ash and maple trees put on a final show before winter. Finally, around the summits, the forest becomes wholly coniferous.

At the higher levels of Huangshan a mighty column of granite towers above the pine forest in a view that is typical of this famous mountain. It is this quintessentially 'Chinese' landscape, often hung with swirling mists, that attracts tourists, mostly Chinese, in huge numbers.

ABOVE Rhesus Macaques (*Macaca mulatta*) have a wide distribution but are scarce in China due to human pressure. The species is most commonly seen among the steep gorges, or in protected parks where it is safe from traps and hunters' guns.

LEFT A young *Sorbus* sapling, hidden in the forest undergrowth on Huangshan's middle slopes, flares in autumnal red against the gloomy background.

Poyang Lake Bird Reserve

On the west side of Poyang the Chinese government have developed a notable bird reserve. During the summer months the entire area is submerged under floodwaters but in winter it becomes exposed and attracts many wintering waterbirds. Farmers are still allowed to cut reeds, graze their buffaloes and rear fish in the various peripheral lakes. These human activities do not detract from the area's suitability for birds; indeed, they are probably essential to ensure the conditions the visiting birds require. There is accommodation at the reserve headquarters in the tiny hamlet of Wuhan, and hardy ornithologists can use hides or walk the banks of the lakes to see a great spectacle of waterfowl sheltering from the cold winds that sweep across the flat plain.

About 200 or more rare Siberian Cranes winter at Poyang lake and, since the demise of the small population that used to winter at Bharatpur in India, these now constitute the entire world population. Wintering geese include the Swan Goose, Bean Goose, Greater White-fronted Goose and Lesser White-fronted Goose. Large flocks of swans keep to the main lake. Oriental Storks, Black-faced Spoonbills, occasional Dalmatian Pelicans and rare Black Storks join the throng, and up to half a million birds may be on Poyang lake at any one time.

The air is filled with the honking of geese and great skeins of them fly overhead in perfect formation. Cormorants and waders add to the multitudes. Some of the cormorants are caught and trained for fishing by the local fishermen, who tie a string around the neck of each bird so that it cannot swallow its catch. Some boats have up to six such cormorants tethered on the bow.

Flocks of Greater White-fronted Geese (*Anser albifrons*) fly low over the reeds of Poyang Lake Bird Reserve to mingle with other waterfowl on the open feeding areas. As the water level drops, more and more of these areas become available to the visiting bird populations. Geese and cranes feed on the leaves and tubers of plants, whilst herons and waders find ample fish and invertebrates to eat.

ABOVE The Swan Goose (*Anser cygnoides*) is a large, long-billed goose that breeds in north-east China, but much of the world population winters on Poyang lake.

BELOW LEFT Though an agile raptor, the Black Kite (*Milvus migrans*) is really more of a scavenger and often dives to pluck refuse from the water surface.

BELOW RIGHT The Baikal Teal (*Anas formosa*) nests in Russia but winters on the Yangtze lakes and the coastal waterways of south and south-east China.

BELOW LEFT The Little Egret (*Egretta garzetta*) is a common flocking bird along lakes, dykes and ricefields.

BELOW CENTRE The Siberian Crane (*Grus leucogeranus*) is one of the world's rarest birds. A few hundred winter on Poyang lake.

BELOW RIGHT The Black-winged Stilt (*Himantopus himantopus*) is a tall pied wader. The pointed wings look batlike in flight.

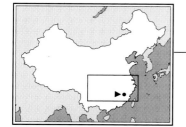

Wuyishan

Wuyishan, at 2,158 metres (7,080 feet) one of the highest peaks in south-east China, is situated on the border between Jiangxi and Fujian provinces on a saddle of rugged mountains that extend through several other provinces. To the north, the Bai Ta and Xin rivers drain all the way to Poyang lake and the Yangtze. To the south, the clear Chongyang river winds through the spectacular scenery of the Wuyigong hills before joining the Min river and heading for the east coast. The region is partly volcanic and partly glacial but the most significant feature is a strong straight faultline that cuts right through the reserve.

This is one of the best developed reserves in China, with facilities to meet any foreign standards. Its headquarters, which include an excellent museum where many of the Wuyishan special species can be seen, are in a beautiful setting on a small stream where Brown Dippers and White-capped Water Redstarts patrol the rounded boulders. Macaque monkeys scamper over rocky screes and noisy groups of long-tailed Red-billed Magpies glide across the valley ahead of the twittering mixed flocks of small tits and warblers.

The road south passes through delightful broadleaf forests and, some 10 kilometres (6 miles) beyond the reserve entrance, reaches the small town of Xingcun. Here, tourists rent bamboo rafts which are poled down the clear stony Nine-bend Creek through a gorge and some amazingly beautiful scenery, with elegant stands of green bamboo standing against backdrops of weird-shaped sandstone rocks and cliffs, eroded by the winds and waters of time and given meaning by the names and legends of the local people.

A sweeping view across Nine-bend Creek and the hills of Wuyishan Scenic Area, seen from the Great King Peak, clearly illustrates the sugar-loaf nature of many of the hills.

ABOVE Seen from Tianyou peak in the heart of Wuyishan Scenic Area, Nine-bend Creek goes through a sharp 180-degree turn around a densely forested sugar-loaf hill. Visitors can enjoy bamboo raft rides down this lovely waterway.

LEFT Wild Pigs (*Sus scrofa*) are nature's refuse cleaners, constantly rummaging through the forest floor vacuuming up any edible litter, fallen fruits, seeds, fungi and carrion, and rooting for worms. In autumn, the pig herds move into the oak zone, feasting on acorns and beechmast to fatten up for the winter.

RIGHT The brightly coloured Fujian Niltava (*Niltava davidi*) is a small flycatcher which keeps to dense thickets. The female, less striking than the male seen here, has a shiny blue neck-patch and white bib but is otherwise brown. The species is quite abundant with a range from the more southerly regions of China to Indochina.

RIGHT The Black-eared Toad (*Bufo melanostictus*) is a common amphibian in woodlands of the moister regions of China. It is a hefty beast, weighing up to a kilogram (over 2 pounds). Also found in gardens, the toads sometimes learn where insects are attracted regularly to artificial lights and emerge each evening in the right places to pick up a good meal of juicy moths.

BELOW LEFT A Chinese stag beetle (*Lucanus*) looks much the same as a European species. The larva lives on rotting wood for several years before the 'antlered' adult emerges. Males sometimes engage in crushing battles for dominance but mostly the great jaws are for show.

BELOW RIGHT A large orb spider (*Nephila*) sits in its huge web waiting for butterflies and other insects to join its larder of tightly wrapped titbits. The male is tiny compared to the female and sometimes ends up being eaten by his mate if he is too amorous.

Yangtze River Gorges

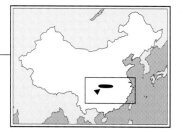

First of the three famous gorges, when heading downriver, is the Qutang, the shortest, at 8 kilometres (5 miles), but most dramatic. Here the Yangtze is compressed through a channel only 100 metres (300 feet) wide to pound through the steep limestone faces that rise to 1,200 metres (4,000 feet) above the river. Climbs up the hillsides near the western mouth of the gorge offer spectacular views into the gorge below and there are fascinating complexes of caves, weird-shaped stalactites and fancifully named rocks and cliffs to help the film flow through the visitor's camera.

The second great gorge is the Wu. It is 40 kilometres (25 miles) long and its sides are so sheer that the sun rarely penetrates to the river below. The Wu gorge is especially famous for its twelve peaks whose dark and sombre grace has inspired the ancient poets and modern visitors alike. These have such evocative names as 'Assembled Cranes Peak', 'Flying Phoenix Peak' and 'Climbing Dragon Peak'. The most famous is the *Shengnu*, or 'Goddess Peak', which supposedly embodies the legendary maiden You Ji who, in pity for the floods and hardships faced by mortals in the area, called upon the great god DaYu to reshape

the mountains by magic powers and allow the turbulent waters to flow smoothly into the Eastern Sea. Finally, after passing the 'Iron Coffin' rock, the visitor emerges from Wu gorge to meet the Flint rapids, where jagged limestone pillars jut from the fast currents to threaten small craft and whirlpools dance menacingly, ever changing their positions with the strength of the flow.

The last of the three gorges, Xiling, is much the longest at 76 kilometres (47 miles) and has always been the most dangerous, with many swirling whirling pools. It is in fact made up of seven smaller gorges and two major sets of rapids. Here, the landscape is the most severe and the most photogenic, marked by famous named rocks, peaks and some ancient monasteries.

BELOW AND OPPOSITE PAGE The Xiling gorge represents the Yangtze river's last headlong plunge from the highlands of western China down to the low-lying plains of the crowded eastern region. Although named as a single gorge, the Xiling is actually a series of gorges alternating with less severe land, allowing for a sprinkling of settlements along the river's banks.

LEFT The sides of the Three Gorges, especially the Qutang and parts of the Wu gorges, consist of spectacular and wholly inaccessible cliffs towering high above the river, shutting out the sun and often trapping swirling mists.

LEFT Less than a hundred of the endangered Baiji or Yangtze River Dolphins (*Lipotes vexillifer*) survive. The species is threatened by dams, pollution and boat traffic. An effort has been made to breed the animals in captivity but, so far, only one has been caught and the project has as yet no chance of success.

BELOW The endemic Yangtze Alligator (*Alligator sinensis*) has been saved from extinction by a successful breeding programme. Several hundred now survive in Anhui province. Though farms are able to breed the animal well, this small species has bony plates in its skin, rendering it much less valuable than other crocodile species which are therefore farmed in preference.

Guangxi Karst Landscape

The province of Guangxi is honeycombed with karst limestone landform, leaving only a small area of the province suitable for agriculture but providing one of the world's most extraordinary and beautiful landscapes.

Probably the most famous and most visited region is Guilin, where 'sugar loaf' limestone pinnacles or *hoodoos* provide a remarkable humped horizon along the bamboo-lined banks of the pretty Li river. Once these hills were a layer of limestone, formed at the bottom of the sea, but this was uplifted through the ages and then eroded by rain and wind. Where small amounts of magnesium are mixed with the limestone it has formed harder dolomite. These hardpoints of dolomite do not erode as fast as the surrounding limestone and become left as nodules which continue to protect the softer limestone beneath, like an umbrella. After a few millennia, the resultant pattern of erosion is what is known as karst. In some areas, like Guilin, pillars are rounded, in other parts of Guangxi the rocky outcrops are more jagged or even left with sharp bladelike edges where it is dangerous to fall. Flatter areas of firmer limestone have sudden pot-holes and underground streams and caves. Much remains unexplored and densely forested. There are many spectacular waterfalls.

Several reserves, including the large area of Damingshan, have been established in Guangxi to protect examples of this particular biotype, which has many endemic and rare specialized plants and animals.

Dusk silhouettes a line of karst limestone pinnacles, reflected in the calm waters of the Tianma river, in Yangshuo county, one of the most visited parts of the Guilin region. Over the years such scenes have almost become synonymous with China.

ABOVE Guangxi limestone forest contains many endemic and rare plants. A high proportion of limestone plants have proven medicinal properties and some species are under pressure from overcollecting, such as the famous 'Dragon's Blood' used to stop bleeding during surgery. Limestone forest is also the home of rare *Trachypithecus francoisi* leaf monkeys and other special fauna.

LEFT Damingshan Nature Reserve is the largest in Guangxi, with dense forests, limestone cliffs and tumbling waterfalls. Much of the area is regrown logged forest but there is also a good area of primary forest with its original fauna and flora intact. Lying in the tropical end of the limestone belt, Damingshan is probably the richest example of this forest type.

RIGHT The Silver Pheasant (*Lophura nycthemera*) is quite common in limestone forests. Each male can gather two or three females. The birds often feed where monkeys and hornbills in the trees above them drop fruit in their own foraging for a meal.

BELOW LEFT Pangolins (*Manis pentadactyla*) are nocturnal insectivores, specialized with their sharp claws for opening up the nests of ants and termites. When alarmed, they curl up in a tight ball, protected by their armourlike scales. Sadly, these scales, which are in fact modified hairs, are highly prized as a medicine by the Chinese and many thousands of pangolins are killed as a result.

ABOVE RIGHT The Rock Python (*Python molurus*) lives in warm forested or scrub habitats. It is a patient predator, waiting near animal trails to pick off unwary mousedeer, rats and even monkeys. It is not as long as its related species the Reticulated Python, but can still reach a considerable size.

RIGHT Most wild bananas (*Musa*) have pendulous flowers pollinated by bats, but this species has a vertical scarlet flower that attracts birds such as spiderhunters and sunbirds to serve as its pollinators. The tiny fruits are quite edible and even the flower is often cooked as a vegetable.

FOCUS ON
TROPICAL SOUTH CHINA

The tropical zone of China is not large but extends in a narrow belt from Tibet to Taiwan and can be divided into a number of distinct sub-units, which include the islands of Taiwan and Hainan. The unit also has the lushest forest and the longest lists of fauna in the entire country.

Tropical areas in the south-east corner of Tibet (Xizang) are part of the south Himalayan system and show great variation in altitude with a huge variety of vegetation types, birds and other faunal groups.

West Yunnan is a land of gorges where the world is bent around the north-east corner of the Indian continental plate. Some 30 million years ago this free-drifting plate finally crashed into the southern belly of continental Asia and subducted beneath that plate, causing a buckling and uplifting of the Asian plate to form the Himalayas. As the Indian plate continued to push northwards, and the Tibetan plateau was raised further, so a series of mountain ranges to the east of the Himalayas became raised and dragged northwards with the movement. These ranges are known in Chinese as the Hengduan Shan or, literally, 'Transversing Mountains'. Squeezed together around the corner of the plate that created the geography of north Yunnan, three of the world's great rivers run parallel to one another with a 70-kilometre (44-mile) span before separating to emerge at sea level, splayed over a distance of 6,000 kilometres (3,700 miles).

The westernmost of these three rivers is the Salween, or Nujiang as it is known in China. This flows west to emerge in southern Myanmar. The second river is the famous Mekong, or Lancangjiang to give its Chinese name, which winds south through southern Yunnan before transversing the Indo-China region to emerge in southern Vietnam. The third is the Yangtze, or Changjiang, which turns eastwards right across China to emerge just north of Shanghai on the east China coast.

This region is very diverse and the three rivers, each flowing rapidly down formidable gorges, act as significant barrier for wildlife. To the east of the Mekong live Black or Concolor Gibbons (*Hylobates concolor*) but on the west, only a few kilometres away, there are White-handed Gibbons (*H. lar*), and across the Nujiang yet another species, the White-browed Gibbon (*H. hoolock*). To the north-east of the Yangtze the Golden Monkey (*Rhinopithecus roxellanae*) occurs, but to the south lives the now very rare grey relative *R. brelichi*, which today survives only on the Fanjingshan mountains of Guizhou province. To the west of the Yangtze but east of the Nujiang lives the black-faced relative *R. bieti*.

OPPOSITE PAGE Rattan canes (*Calamus*) send out their spiny-rayed leaves as they clamber to reach the forest canopy of tropical southern China. The strong pliable canes are used to make furniture and, with demand exceeding growing stock, many rattan species are becoming rare as a result of overharvesting.

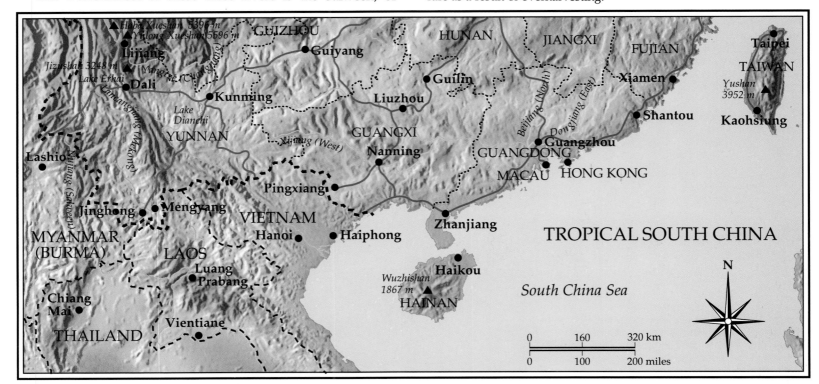

ABOVE The Elephant-eared Yam (*Alocasia macrorhiza*) spreads its huge leaves in the forest undergrowth. The sap of the leaves and stem is an irritant but is used locally for medicine.

LEFT China is a land of bamboos – from the giant species of the south to the ornate, plump-noded Buddha bamboo and the slender dwarf arrow bamboos of the mountain tops. This tropical species is *Dendrocalamus strictus*, often used for food.

OPPOSITE PAGE Strangling figs (*Ficus*) grow into monstrous and complex shapes as descending rootlets find the ground and swell up into new supports and buttresses. The sap of the cut fig flows like white blood, arousing superstitious ideas that these plants have feelings. Most ethnic groups in South-east Asia have a taboo against cutting down figs.

Pheasants and many plants show similarly complex local distributions, but some animals, like the strange goat-antelopes the Takin, Serow and Goral, have easily traversed the whole region.

The south of Yunnan province is a lowland area with dense rainforests, especially in Xishuangbanna, known in China as the 'kingdom of plants'. Here China's last herds of elephants roam and Gaur, Tiger and Green Peafowl give the region a truly tropical flavour.

Xishuangbanna is a Dai minority autonomous prefecture, tucked in between Laos and Myanmar. The name itself is taken from the Dai-Thai name *Sip Song Panna*, or 'twelve districts'. It is very much a frontier state and the population is mostly made up of minority groups – River Dai, Black Dai, Jinuo, Bulang, Yao, Hani and several other groups, each with their own language. Some, like the Dai, have their own script.

For centuries the Chinese rulers were wary of Xishuangbanna. It was a wild land of dangerous animals, snakes, diseases and hostile villagers. Those posted to the region undertook family funerals before their move, just in case. Today the region is less wild. The different ethnic groups are very friendly, roads have opened up the countryside, and malaria and other diseases have been controlled. Many Chinese have moved into the area to grow rubber and a fast-expanding tourism industry is the new money-spinner.

Xishuangbanna is China's biggest rubber-growing region, and the clearance of rainforests here to make room for rubber plantations has been a biodiversity tragedy. Many of the richest forests have been lost in the process and the economic benefits are highly dubious. In the days of China's economic isolation it may have made sense to be self-sufficient in natural rubber but in an open economy it is pointless to spread a crop for which there is already a global surplus. In addition, China is too far north to attain good rubber yields and the trees, which are evergreen in equatorial regions, lie bare for several winter months. Moreover, the plantations provide poor soil protection and are characterized by serious levels of soil erosion. Also, when the latex is harvested it requires heat-processing, which involves the cutting of large quantities of natural firewood in these areas that lack coal or alternative fuel.

However, these developments have not totally replaced the tropical forests, which still cover a quarter of the prefecture area. Five nature reserves have been established – Mengyang, Menggao, Menglun, Mengla and Shangyong.

The tropical zone of China is rich in reptiles. This large agamid lizard, *Physignathus cocincinus*, looks like a South American iguana. It lives along streams in dense forest but is very agile and climbs high into the tree canopy when alarmed or threatened.

The Rock Python (*Python molurus*) is a large, fat constricting snake that hunts at night by using its keen sense of smell. It kills its small mammal prey first by striking and then by crushing, before locking the victim in its large mouth full of sharp, backward-pointing teeth.

In Mengyang, on a fine spring morning, the air is alive with the cries of the Great Barbet (*Megalaima virens*), whose endless *miaow* calls sound like an army of angry cats. Splendid Greater Racket-tailed Drongos (*Dicrurus paradiseus*) sing and cackle in the upper canopy and the amazing shiny blue of the red-eyed Asian Fairy-bluebird (*Irena puella*) flashes through the leaves. At ground level, Junglefowl (*Gallus gallus*) and long-tailed Silver Pheasants (*Lophura nycthemera*) patrol their territories and parties of White-crested Laughingthrushes (*Garrulax leucolophus*) send out their demonic laughter. Along the streams the Slaty-backed Forktails (*Enicurus schistaceus*) flicker their banded, black-and-white tails whilst the rare Blyth's Kingfisher (*Alcedo hercules*) sits silently on a favoured perch before diving into a quiet pool to emerge with a small wriggling fish.

Leeches loop through the leaf-litter – sniffing out holes in tourists' tennis shoes and causing squeals of horror from parties of Chinese schoolgirls. Giant orb spiders hang menacingly in their sticky golden webs, waiting for courtship-dazzled butterflies to take an incautious route. A pile of steaming fresh elephant dung heaves mysteriously above the excavations of huge scarab beetles and the wet patch where an elephant has urinated attracts a flitting carpet of beautiful swallowtail and yellow butterflies.

But the rains are not far off and soon the sound of thunder and the heavy pounding of rain will silence the barbets and the clear streams will swirl with muddy run-off. The elephants will relish the cool shower in their bamboo thickets but the kingfisher will have to work hard to find enough fish for hungry fledglings in the dark hole under the old fig tree.

Menglun town is the site of the famous Menglun Botanical Institute and Botanical Gardens – a beautiful spot in a bend in the Xiaoheijiang river where there is an excellent collection of the rare plants of the region, including an innovative *ex-situ* conservation garden where many rare species are propagated inside an existing block of old logged forest. The botanic gardens are connected to the small town by a spectacular cable footbridge across the river. In the evenings, the Dai girls bathe their long tresses in the river whilst a fisherman working from a bamboo raft tries to catch the 2-metre (6½-foot) long Mekong Catfish (*Pangasius sanitwangsei*). As dusk falls the rare Giant Flying Squirrels (*Petaurista petaurista*) come out of their tree holes, rest for a while in the rays of the setting sun, rump their way clumsily to the end of a tall branch then launch off in amazing glided flight into the dark forest to find their food for another night.

Hainan Island is another important part of China's tropical heritage and contains some of its best forests, although only 7 per cent of the original lush cover remains. Most of the forest in the north and east has been totally replaced by farmland, urban

The tropical lantern bugs are related to cicadas and, like them, feed by sucking plant juices. The function of the characteristic bulbous 'lantern', which varies greatly from species to species, is largely for species recognition and attraction of mates.

The Common Tiger Butterfly (*Danaus genutia*) is distasteful to birds and is mimicked as a protective measure by other less obnoxious but generally rarer butterflies. The eggs are laid on the toxic milkweed plants.

development and extensive plantations of rubber, tea and eucalyptus, while titanium mines have opened up other areas.

In the north-east corner of this island province there are some important mangrove forests at Dongzhaigang and Qinlangang, but the more important dryland forests are in the hilly country of the south-west. The tallest peak and largest reserve on Hainan is Wuzhishan, or 'Five-finger Mountain', which rises to 1,867 metres (6,125 feet). This mountain has the widest range of altitudinal zones found on the island and is the collecting site of many of Hainan's special plants and animals such as the endemic Hainan Partridge (*Arborophila ardens*), the rare White-eared Night-Heron (*Gorsachius magnificus*) and Blyth's Kingfisher (*Alcedo hercules*).

However, farmers and loggers have already encroached to rather high altitudes, and although the mountain is splendidly scenic and an excellent reservoir for Hainan's montane flora and fauna, it is no longer the best site to see lowland forests. For that, the place to visit is the small reserve of Bawangling, where Hainan's last gibbons still sing their plaintive morning chorus and the Hainan Partridge is still common. Hainan's own races of Grey Peacock-Pheasant (*Polyplectron bicalcaratum*) and Yellow-billed Nuthatch (*Sitta solangiae*) are both also common in Bawangling. Giant *Dacrydium* trees grow along the highest ridges of this beautiful reserve, testifying to its ancient and undisturbed condition.

The forest here is more rich in palms than any other in China. The ground is covered in an undergrowth of *Licuala* palms whilst tall fanpalms (*Livistona*) rise into the canopy; *Caryota* palms display their characteristic fishtail-shaped leaves, small *Pinanga* palms occupy the sapling layer and several species of climbing rattans (*Calamus*) clamber over other trees to reach the sunlight of the upper storeys.

The east and south coasts of Hainan are regularly buffeted by typhoons, which prevent the forest there from reaching great height, and have a high rainfall, but the north-west coast is in a rain shadow and is much drier. This corner is also a fast-draining sandy area where a more park-like deciduous forest is typical and where natural grasslands form a savannah ecosystem. Here are two of Hainan's important mammals – the endemic hare, *Lepus hainana*, and a local race of Eld's Deer (*Cervus eldi hainana*).

Hainan has a tropical coastline with beautiful beaches,

splendid rounded granite rocks, a scrubby beach vegetation and some excellent coral gardens. Sadly, uncontrolled tourism has spoilt much of the scenery, and dynamiting and mining of coral for making cement and lime has destroyed or damaged many of these reefs. Some good diving areas can still be enjoyed along the north coast and other areas may recover when stricter controls can be put in place.

Hainan has one of the fastest economic growth rates in China but the local government are conscious of the natural values that they are in danger of losing and are making a serious plan to incorporate nature conservation in the development process. A special council has been established for 'The Development of Hainan in Harmony with the Natural Environment'. We must hope for the success of this initiative if the fragile ecosystems of this beautiful island are to survive.

The southern fringes of Guizhou, Guangdong and Zhejiang provinces, and the small territory of Hong Kong, are also tropical. This zone is of moderate richness but is also affected by typhoons and a markedly seasonal climate. Only a few relatively pristine areas remain but some of the secondary forests are full of wildlife and the coastal areas are extremely important for migrating birds.

Despite its high population density and advanced development, Hong Kong preserves some remarkable wild places and valuable biodiversity. The territory boasts a list of over 500 bird species and has retained or regained a healthy fauna of invertebrates, amphibia and reptiles. This is all the more surprising when one realizes that the entire territory was originally deforested by typhoons and man-induced fires. When the island was first colonized by the British the hills were almost bare but the colonial administration set about a programme of fire control and replanting, so that by 1940 extensive areas were under secondary forest. During the Second World War the occupying Japanese cleared most of these forests but after the war the programme of reforestation was renewed and over 25 per cent of the territory was included in an extensive system of country parks.

Some animals have colonized Hong Kong through introduction. The only squirrels in the territory are the Red-bellied Squirrels (*Callosciurus erythraceus*) a form from south-west China and Thailand but now very common on the island. The

ABOVE The Collared Scops Owl (*Otus bakkamoena*) is common in the woodlands and scrubland of east and southern China. It feeds on small vertebrates but mostly takes insects, hunting from a low perch at night.

BELOW The anemone-like Fanworm (*Spirobranchus*) of tropical China's coral reefs spreads its filter-feeding arms to sift through passing debris and draw edible plankton or food particles to its central mouth. When resting, the worm withdraws into a hole in the coral.

natural monkeys are the Rhesus Monkeys (*Macaca mulatta*), but there are also troops of the smaller Long-tailed Macaques (*M. fascicularis*) which naturally occur much further south and are absent from south China. There are even some hybrid groups where Long-tailed Macaques have become mixed up with Rhesus. In the Central area of Hong Kong visitors are amazed to meet Indonesian Sulphur-crested Cockatoos (*Cacatua sulphurea*), Rainbow Lorikeets (*Trichoglossus haematodus*) and other exotic birds flying freely and breeding successfully.

Taiwan is another biological treasurehouse. With an area of 35,760 square kilometres (13,800 square miles) this is the largest island off China's coast. It lies 130 kilometres (80 miles) from the mainland at the narrowest point of the Taiwan Straits and, having been separated for many thousands of years, has acquired a long list of endemic species. The southern areas of the island are truly tropical but the central and northern portions are subtropical, with temperate and subalpine zones on the main mountain chain which stretches the entire length of the island.

Along the western shore the sea is shallow, contours are gentle and there are extensive mudflats and stunted northern mangroves. On the east coast, however, the mountains plunge steeply into the sea. The sea profile continues to plunge and not far from the coast is a deep sea trench that goes down some 6,000 metres (20,000 feet). The mountains themselves are geologically very young and still rising two millimetres per year. There are frequent landslides and earthquakes in this unstable zone.

The highest peaks of Taiwan rise to over 3,000 metres (9,800 feet), crowned at 3,952 metres (12,966 feet) on the summit of Yushan. The mountain forests are coniferous but lowland forests, especially in the south of the island, are of a typically tropical nature with many palms and lianas. The ring of tropical evergreen monsoon forests that once encircled the central mountain chain has now largely been lost, though some patches still survive in the far south.

Fauna on Taiwan contains such endemic notables as the Taiwan Black Bear (*Ursus thibetanus*), Taiwan Macaque (*Macaca cyclopis*), Formosan Serow (*Capricornis crispus*), several small mammals and at least 14 species of birds, including two endemic pheasants, a partridge and a spectacular blue magpie. Despite its large human population, Taiwan has been able to preserve over 50 per cent of its natural forest cover – a higher proportion than any other Chinese province.

The coastline of tropical south China still has some valuable coral reefs, though these are seriously threatened by harvesting of coral for making lime and cement as well as land-borne pollution and sedimentation. On some of the islands, and on a few beaches, marine turtles still come to nest – smaller Hawksbills and large Green – often falling foul of fishermen's nets and ending up as soup or stuffed curios in the tourist stalls. Shoals of playful dolphins frolic with passing boats and in Hong Kong waters the rare Chinese White Dolphins (*Sousa chinensis*) still persist, despite the high levels of water pollution.

In sheltered bays along the south coast, on Hainan Island and at Hong Kong's famous Mai Po marshes can be found mangrove forests. Close to Hong Kong's border with China the land is swampy and there are many shrimp and fish ponds (traditional tidal forms of which are known as *geiwai*). Mangroves grow in the coastal creeks where many egrets, herons and waders find their livelihood. Some of the *feng-shui* or 'good luck' wooded hills act as rookeries for these waterbirds but the real concentration area for waterfowl is the Mai Po marshes, where WWF-Hong Kong have established an excellent nature reserve.

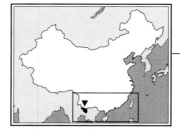

Xishuangbanna Nature Reserves

Of Xishuangbanna's five nature reserves, the largest and best place to see China's last herds of elephants is Mengyang where many large-tusked bulls and the scattered families of cows and young wander between salt licks. Gaur, a large species of wild cattle, also live here and make regular visits to the salt and mud wallows. Leopards stalk wild pigs, sambar deer and macaque monkeys, and hornbills still clatter through the canopy, chasing crickets with their unwieldly bills.

Menglun is a smaller reserve, divided into three blocks and distinguished by being composed mostly of white limestone. Huge-buttressed *Tetrameles nudiflora* trees tower beside jagged karst cliffs and pillars, and bats flit in and out of small caves. Elephants are rare in Menglun but there is a good herd of Gaur and the forest floor is ploughed up by their sharp hoofprints.

Menggao is the only reserve to the west of the Lancangjiang river, famous for its natural pine forests. Mengla and Shangyong are more remote and situated along the Laos border. They are very wild and still have tiger, Gaur, elephants, gibbons, Great Hornbill and leaf monkeys. Here also grow the tallest trees in all China – the dipterocarp forests that dominate the vegetation of South-east Asia being represented by small patches of *Parashorea chinensis*, whose straight trunks rise 30 metres (100 feet) before branching out into compact round crowns. At Mengla, a walkway has been constructed where visitors and scientists can walk high above the ground through the canopy to get a closer look at the canopy wildlife and to study the small mammal populations of lorises, squirrels, tree shrews and arboreal rats.

A clear stream in the depths of the forest in Mengyang Nature Reserve lies still and half empty near the end of the dry season. Red wild figs, fallen from a tree above, lie rotting in the water, filling the air with a pungent aroma.

This climbing *Rhaphidophora decursia* cheeseplant is one of many climbers and lianas that festoon the trees of a typical tropical rainforest, hitching a lift to the light at the top of the canopy. The species is much admired as a houseplant because of its tolerance of growing in shady places.

Herds of Gaur (*Bos gaurus*) roam the forests of Xishuangbanna, feeding on the grasses beneath the subtropical oaks and visiting mineral salt licks to obtain vital nutrients. They can occur in secondary forest and may range up to 1,400 metres (4,600 feet) in altitude. These wild cattle are very shy and local hunters take great pride if they can kill one with their primitive muzzle-loader muskets.

ABOVE The dazzling male Fairy-blu... ...a *puella*) is one of the jewels of Xishua... forest canopy. It is a common visitor to...e fruiting fig trees. The less gaudy femal... is a duller greenish blue.

RIGHT The Long-tailed Broadbill (*Psarisomus dalhousiae*) nests over streambeds and often travels in small parties. The snap of its powerful bill as it catches large insects in the forest canopy is loud enough to be heard from many metres away. The species has a loud whistled call.

The *Creoboter* mantis has very fast reactions and can snatch passing butterflies out of the air as they fly past the everwaiting arms. Its wings are decorated with curious black and white false eye-spots.

The Sun Spider's (*Gastercantha*) hard shell and extraordinary tail-like appendages make it difficult for wasps or birds to eat. With this protection it can hang boldly all day guarding its web.

A colourful clump of *Impatiens aquatica* decorates the middle of the Sanchahe stream in Mengyang Nature Reserve.

The orchid-like flower of *Eranthenium* species, a fairly common plant of the forest floor in Mengyang.

Flowers of *Thladiantha* species. This small climber is a wild relative of the cucumber.

The Basket Fungus (*Dictyophora indusiata*) smells of rotting meat, which attracts flies to its slimy cap.

The strange swallowtail butterfly *Lamproctera curius* has transparent wings and looks like a dragonfly. These insects gather in large swarms to sip at mineral-rich streams.

This nymphalid butterfly (*Parathyma*) copies the colour pattern of black and white bands commonly exhibited by many *Neptis* species that are distasteful to predators.

The beautiful Paris Peacock Butterfly (*Papilio paris*) enjoys a cool sip of elephant urine alongside a stream in the Mengyang reserve.

The Red Lacewing (*Cethosia biblis*) is another butterfly which gains some protection by imitating the colour pattern shown by many distasteful species.

The Leaf Butterfly (*Kallima inachus*) is beautifully camouflaged when its wings are closed, but when they are opened in courtship a dazzling flash of orange and iridescent purple is revealed.

The Great Mormon Swallowtail (*Papilio memnon*) is one of the largest swallowtail butterflies in China. The larva feeds on grapefruit bushes and can become a pest in plantations.

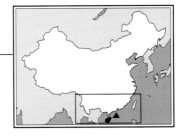

Hainan Island Nature Reserves

The moist forests of south-west Hainan are an important biological repository. The tall mountain of Wuzhishan is the largest reserve on the island and best place to examine the montane flora and fauna, whilst the pretty lower reserve of Bawangling preserves the richest mid-altitude forests and endemic gibbons and partridges.

Further south-west, in a slightly drier part of the island, is Jianfengling where a core nature reserve is bordered by a large buffer of logged forests and valuable botanic gardens. The primary forest is lush, full of palms, wild lychee trees and lianas, and alive with birds and reptiles. Gibbons are absent from Jianfengling but there are still many Rhesus Monkeys, for which a special reserve is maintained on the south coast at Nanwan. Here busloads of tourists buy nuts and fruits to throw to the several hundred habituated monkeys that rush to fight and squabble over these titbits. Less daring juveniles sit wistfully in neighbouring trees, wishing they were strong enough to compete for the easy food and watching for some overlooked morsel that they can quickly salvage when the bigger monkeys at last have moved on.

The Hainan race of Eld's Deer became so rare that the last wild animals were rounded up at the Datian reserve and kept in a large fenced enclosure in semi-wild conditions. They now flourish and the population has built up to a healthy number once again. The problem today is that little natural habitat remains for the species and there is still no control on hunting on the island. So, for the time being, the animals are safest where they are and new holding farms are being established at both Datian and Bangxi.

Large treeferns (*Cyathea spinulosa*) add a tropical and primitive atmosphere to the lush rainforests of west Hainan. They have hard, spiny trunks and a beautifully balanced and symmetrical leaf structure.

ABOVE The huge twisting roots of a giant banyan (*Ficus*). A tree of this size provides great quantities of small fruits for canopy animals such as pigeons, bulbuls, primates and squirrels. Local people often give offerings and say prayers at the base of such old trees, which they believe can develop a soul or spirit.

RIGHT Trees and ferns in the uncut core zone of Jianfengling Nature Reserve. Much of the outer perimeter of the reserve was formerly within a timber production unit and is less pristine. However, even this can recover and be recolonized from the core zone.

ABOVE Eld's Deer (*Cervus eldi*) have become extinct in south-west China and Vietnam but still persist in Hainan where there is an insular race (*C. e. hainana*). The last wild animals were rounded up and kept in a large fenced enclosure at Datian reserve. This captive herd has flourished and now numbers about 500. It may, in the future, be possible to reintroduce these back into the wild.

LEFT The ubiquitous Rhesus Macaques (*Macaca mulatta*) crop up in rugged terrain all across southern China. They live in boisterous groups and can become quite tame if not persecuted. Hainan has set up some special reserves for these monkeys where hand-outs from visitors have enabled populations to reach high densities.

In repose this small *Yasoda tripunctata* butterfly is inconspicuous and the false antennae would anyway draw a bird's attention to the rear of the insect. When open, the upper wings are a brilliant blue.

The Faun Butterfly (*Faunis eumeus*) is a beautiful member of the Amathusid family. It is found only in shady moist forests throughout China's tropical zone, including Hong Kong.

The White-tailed Robin (*Cinclidium leucurum*) is a shy bird, keeping to the densest thickets of the forest. When alarmed, it flicks the white flags of its tail whilst giving a *click click* call.

Chestnut Bulbuls (*Hypsipetes castonatus*) are common in the forest canopy of southern China and Hainan. They feed in fruiting trees and bushes and sometimes join mixed species flocks.

A pattern of strange concentric rings indicates the work of a leaf-eating beetle. By keeping to a tight spiral the insect minimizes the danger of being seen moving and always meets fresh undamaged leaf.

The leaves of the *Aristolochia* vine contain a poisonous latex. Some rare butterflies, such as *Troides* and *Atrophaneura*, lay their eggs here and the larvae also become poisonous, and thus protected, by eating the leaves.

ABOVE The white flowers of jewel orchids (*Macodes*) glow on the dark forest floor of Jianfengling Nature Reserve.

RIGHT *Selaginella* is a common primitive fern that grows as a ground cover in moist forest areas. The species is a good indicator of evergreen condition as it is rather intolerant of desiccation. The leaves are used by local villagers as sterile dressings on wounds and to stop bleeding.

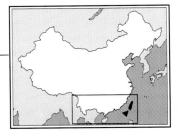

Taiwan Island

Despite a huge human population, Taiwan retains over 50 per cent of its forest cover. Most of these forests are coniferous and found along the main mountain chain, but there are lowland forests along the east coast and in the extreme north and south of the island. At Kenting National Park, in the south, the forests are tropical, dominated by *Myristica cagayensis*, *Pterospermum niveum* and *Artocarpus lanceolatus*. Great banyan trees are characteristic. There are coral reefs along the coast.

In the north there are extensive regions of limestone, characterized by steep cliffs and deep ravines with a specialized alkali-tolerant sparse forest. The extreme north is typically subtropical, with lowland forests dominated by oaks and chestnuts. Climate also varies: in the south the rainfall is mostly in the typhoon season between June and September, but in the north there is more rain in the winter months.

Faunally, Taiwan matches the subtropical zone of mainland China but the island has been isolated for millennia and there has been a considerable degree of local evolution. Many endemic species and subspecies occur, such as Formosan monkeys, Formosan Serow, Formosan Black Bear, two endemic pheasants, an endemic sibia, a laughingthrush, a magpie and a tit. There are many nature reserves in which to enjoy these local specialities, and there is a generally high standard of conservation awareness.

Yushan, at 3,952 metres (12,966 feet), is Taiwan's highest mountain as well as the focus for Yushan National Park, largest of the island's five national parks. It is also the wildest, with deep valleys and 30 peaks over 3,000 metres (10,000 feet) high providing a vast range of habitats from subtropical forests to alpine grasslands and the bare rock of the highest summits.

Taiwan is well endowed with waterfalls. ABOVE Tsaihung waterfall, on the northern edge of Yushan National Park, is one of the region's most beautiful. RIGHT Wufengchi, in Ilan county, is a massive three-level fall (only the upper fall is shown here) cascading through dense subtropical forest.

BELOW At Alishan, a mountain resort on the edge of Yushan National Park, massive stumps of 2,000-year-old Taiwan cypresses, now surrounded by a young forest, stand as testament to the ancient forest that was once here.

ABOVE A view of the rugged terrain of Yushan National Park, capped by the snow-covered peak of Yushan itself.

OPPOSITE PAGE The north-east coast of Taiwan is characterized by fearsome cliffs and rocks that plunge precipitously into the depths of the Pacific Ocean.

ABOVE Two damselflies (order Odonata) joined in copulation hang limply waiting for the sunshine before resuming their nuptial flight.

LEFT The Taiwan Black Bear (*Ursus thibetanus*) differs only slightly from the mainland race. The species, which lives in the oak forests of the main mountain range, is protected.

BELOW LEFT Taiwan has an endemic Serow (*Capricornis crispus*). Serow live in steep rocky terrain and are extremely agile. When resting they tend to sit on prominent rocks from which they can keep a good view of their surroundings and pick up the sound of any approaching predators with their large sensitive ears. If danger threatens, they will take shelter in crevices.

BELOW RIGHT The Red-bellied Squirrel (*Callosciurus erythraceus*) is common, with several races across southern China. It feeds on flowerheads and fruits and nests in tree holes or twiggy dreys.

ABOVE The Taiwan Yuhina (*Yuhina brunneiceps*), seen here clinging to a branch of cherry blossom, is a small tit-like species, endemic to the island and quite common.

RIGHT The Yellow Bittern (*Ixobrychus sinensis*) frequents the reedbeds and ricefields of Taiwan, eastern and southern China. It is often a victim of trappers with mist nests and is sold for food.

BELOW The Mikado Pheasant (*Syrmaticus mikado*) is one of Taiwan's two endemic pheasant species. The pointed tail is barred black and white and the body feathers are edged with purple. This rare bird lives on the main mountain chain above 1,800 metres (5,900 feet).

BELOW Another endemic bird is the beautiful and noisy White-eared Sibia (*Heterophasia auricularis*). It is decorated by delicate elongated filamentous plumes on the ear coverts. Sibias, which are not uncommon, often gather in small parties and call with a resonant rising *fei fei fei*.

LEFT Treeferns (*Cyathea spinulosa*) growing in Yangmingshan National Park. The species is endangered because its complex fibrous stems are ideal strata for cultivating exotic orchids.

BELOW RIGHT Pink *Hedychium* flowers in Alishan, Taiwan.

Yushan National Park is one of many areas on Taiwan where conservation measures have preserved the island's original vegetation. ABOVE LEFT The open scrub provides habitat for many pretty flowers, among them this *Dianthus* species. ABOVE CENTRE The white florets of *Aensliaea* are traditionally presented as a gift to win the love of the recipient. LEFT In the ever-humid spray of a waterfall grow a wealth of small plants such as these *Adiantum* ferns.

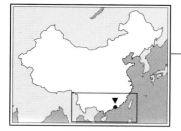

Hong Kong Territory

As Hong Kong's forests have grown after its almost total clearance at some time in the past, so wildlife has returned. With forests being cleared quickly in adjacent provinces of southern China, birds and other creatures were only too ready to colonize the recreated habitat. Every year a few new species are recorded for the territory.

The best rainforest area is at Taipo Kau in the New Territories. Here natural forest patches, secondary forests and plantations form a valley catchment which is now a nature reserve. Visitors are surprised by the wealth of wildlife. Rhesus Monkeys scamper through the canopy, where a lot of interesting forest birds can be found – Velvet-fronted Nuthatch, Oriental Turtle-dove, Japanese White-eye, Fork-tailed Sunbird, Grey-chinned Minivet, and many more. The night watcher may even see wild pigs, porcupines, leopard cat, palm civet and a variety of snakes.

Even on Hong Kong Island there is now extensive forest, especially around Victoria Peak, which provides a semblance of what Hong Kong must have looked like hundreds of years ago. Graceful *Livistona* palms mingle with broadleaf trees and occasional gnarled figs, whilst the undergrowth is clothed with saplings, great-leaved wild yams and colourful-flowering wild *Hibiscus* bushes. The Blue Whistling-Thrush rustles the leaves of the forest floor in search of food. Gaudy black and yellow *Troides* butterflies glide over the canopy and a myriad insects and reptiles go about their daily lives quite unaware that 100 years ago the hill was bare.

The coast of Lamma Island, one of Hong Kong's outlying islands, looking southwards towards its highest peak, Mount Stenhouse. Many of these outer islands are characterized by rugged coastlines alternating with beautiful sandy beaches.

RIGHT One of the wildest views in all Hong Kong: the isolated eastern coast of the New Territories, much of it protected within Pat Sin Leng Country Park. Many of the offshore islands are ringed by corals, soon to receive protection as marine parks.

BELOW Much of Lamma Island consists of fire-maintained grasslands, but there are also numerous *feng-shui* woodlands and, as seen in the foreground, woodlands formed both by plantation and natural regeneration.

OPPOSITE PAGE, BELOW A crystal clear pool in a mountain stream on Lantau Island. Though now site of Hong Kong's new airport, Lantau remains one of the wildest parts of the territory. Its mountain streams are ideal habitat for such endemic species as the Hong Kong Newt and Romer's Tree Frog.

ABOVE The rocky coastline of Lamma Island. Much of Hong Kong's coast, as well as that of nearby parts of Guangdong, is characterized by massive outcrops of rounded granite boulders, forming low cliffs and offshore islets.

BELOW The Rufous Fantailed Warbler (*Cisticola juncidis*) is a common inhabitant of ricefields and reedbeds in southern China. However, it is difficult to approach and hides in the reeds, soaring into the air again some distance away with a fast *zitting* call.

BELOW The Red-whiskered Bulbul (*Pycnonotus jocosus*) is one of Hong Kong's commonest birds, favouring gardens, parks and open scrub. Only the adult has the whiskers that give the species its common name, though juveniles do have the long crest.

ABOVE LEFT The tiny Romer's Tree Frog (*Philautus romeri*) looks unremarkable but it is Hong Kong's only endemic frog, confined to three or four localities including the site of the new airport. A conservation project is under way to save the species by capture, captive breeding and release into new sites that seem to be suitable habitat.

ABOVE RIGHT The Hong Kong Newt (*Paramesotriton hongkongensis*) is endemic to the territory and adjacent parts of south China. It lives in small streams and ponds, feeding on worms and occasional larvae and laying its single eggs under the leaves of waterplants. The juveniles stay in the water and develop long feathery gills for direct respiration. These gills are lost with maturity when the newt has to rise to the water surface to gulp air into its lungs.

RIGHT This grass lizard (*Takydromus sp.*) is an active insectivore of rocky and wild places. Although it may sometimes look as though it is sleeping in the sunshine, it is always alert and dashes off with great speed when approached or threatened. As with other lizards, it can lose its tail to a predator and grow another almost as good.

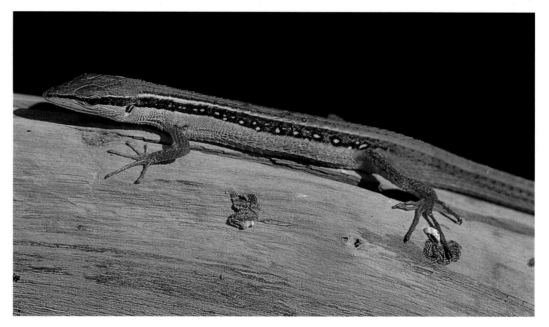

RIGHT A common snake of moist areas in southern China, the Red Keelback (*Rhabdophis subminiatus*) is inclined to give a venomous bite if handled. It can slither very fast and catches frogs and other small vertebrates, which it eats whole and head first.

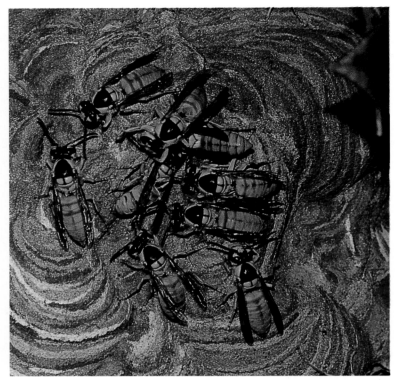

ABOVE LEFT A newly hatched Brown Cicada (*Cryptotympana mimica*). The larval cicadas live under the ground, feeding off plant roots, but the adults suck the sap from under the bark of trees through the long piercing mouthpiece.

ABOVE RIGHT Aggressive paper wasps (*Vespa*) make elaborate nests from regurgitated chewed wood. A single queen lays eggs in each cell of the nest to be tended and guarded fiercely by the worker wasps.

LEFT The cicada *Scieroptera sanguinea* lives in warm forests. Its bright red and black coloration is a warning to birds that the insect is distasteful.

BELOW LEFT The startling blue berries of *Acanthopanax trifoliatus* shine beside a forest trail in the valley of Tai Po Kau. The roots of this plant are used by the southern Chinese as a substitute for ginseng in health tonics. Ginseng itself is found only in the north.

OPPOSITE PAGE, ABOVE The brightly coloured tigermoth *Dysplania militaris* uses its wing pattern as an advertisement that it is not palatable to predators. Thus protected, it can fly safely in the daytime, even during copulation.

OPPOSITE PAGE, BELOW LEFT The coral-red flowers of *Erythrina stricta* are among the most dazzling in the southern forests. They attract many birds, which act as the main pollination agents, and it is worth waiting at such trees to get good views of the local avifauna.

OPPOSITE PAGE, BELOW RIGHT Another insect displaying bright warning coloration, a female birdwing butterfly (*Troides aeacus*) lays her eggs on the poisonous *Aristolochia* vines that will protect both larva and eventual adult from bird predation because of the toxic intake.

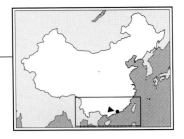

Mai Po Marshes

At Mai Po, clever landscaping of pre-existing fishponds by WWF-Hong Kong has resulted in some raised islands, deeper ponds, other tidal areas, mangroves and reedbeeds, which all provide a wide range of wetland habitat for birds. The birds have responded. They recognize the safety of Mai Po and they visit in ever-growing numbers.

Many species such as Little Grebe, Common Moorhen, Common and White-throated Kingfisher, Little Egret, Grey Heron, and White-breasted Waterhen can be seen all year round. But many others are winter migrants, using Mai Po as a safe wintering area or staging post for journeys further south. Pied Avocets, Common Coots, Common Black-headed Gulls, the rare Saunders's Gull, Dalmatian Pelicans and Oriental Storks, huge rafts of Northern Shovelers, Common Teal, Falcated Duck, Mallard and Tufted Duck, fair numbers of Black-capped Kingfishers and great flocks of sandpipers and other waders are among these shoreline visitors, whilst tiny Prinias and other warblers live among the reeds.

A huge population of cormorants has built up at Mai Po, originally coming as winter visitors but a few now staying all year round. These take quite a few fish from the fish farms and are unpopular with local residents but in China the species is getting rather rare, so its protection is important.

Birds of prey are attracted to so much avian activity. Scavenging Black Kites and a few fish-catching Ospreys are regular residents, but the larger eagles that visit Mai Po are seasonal.

WWF has developed an excellent system of walkways, raised hides, blinds and even a floating catwalk onto the mudflats so that birdwatchers can get good views. There is a well-equipped education and training centre.

The mangrove vegetation is also important. Mangroves along south China have been mostly destroyed, so Mai Po's significance for conservation grows. Dominant species are *Kandelia candel* and *Avicennia marina*, but the holly-leaved herbs *Acanthus ilicifolius* are also an important zone of the mangrove system. With the mangroves comes a wide range of other wildlife – snakes, pop-eyed mudskipper fish and armies of busy fiddler crabs. Occasionally otters and leopard cats are seen on the marshes.

BELOW The outermost edge of the mangroves at Mai Po. OPPOSITE PAGE, ABOVE A view across one of the *geiwai*, or tidal shrimp ponds. OPPOSITE PAGE, BELOW Mai Po provides a wide range of wetland habitats for resident and wintering waterfowl.

ABOVE, LEFT AND RIGHT The Chinese Pond Heron (*Ardeola bacchus*) is a handsome resident of southern China's ricefields and coasts. In winter, the bird is drab and grey but in summer it has a rich chestnut mantle and plumed back. As it leaps up in flight the white wings spread in a startling flash.

LEFT Mai Po is an important wintering area for the rare Saunders's Gull (*Larus saundersi*) which is endemic to the east coast of China and breeds only in two small localities.

BELOW LEFT The Sharp-tailed Sandpiper (*Calidris acuminata*) is a rare migrant on the east coast of China, favouring mudflats and mingling with other waders.

BELOW RIGHT The dainty Lesser Sandplover (*Charadrius mongolus*) is frequently seen among Mai Po's throngs of waders.

ABOVE During the daytime Night Herons (*Nycticorax nycticorax*) hide up in large colonies in the dense mangrove thickets. At dusk, they emerge with harsh croaked calls to fly off to open feeding areas.

ABOVE The White-throated Kingfisher (*Halcyon smyrnensis*) is both a resident and a winter visitor in Mai Po. It takes fish from the *geiwais* but also catches insects and lizards on land.

BELOW The Terek Sandpiper (*Tringa cinerea*) is a common passage migrant to the south China mudflats. It has a distinctive upturned bill and dark wings against a greyish body.

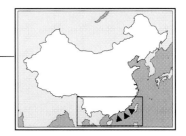

South China Marine Life

The south coast of China is tropical and, in addition, China's seas extend far south to include the tropical archipelagos of Xisha, Zhongsha and Nansha, though some of this marine territory is disputed by neighbouring countries.

These areas are characterized by extensive coral reefs dominated by hard corals such as the branching *Acropora*, brain corals such as *Platygyra* and soft corals like *Dendronephthya*. Among the filamentous sea anemones, colourful clownfish play peek-a-boo with molesting human or piscian observers, impervious to the stinging nematophores in their hosts' tentacles. Elegant butterflyfish flit among the corals in small shoals, whilst the strong-beaked parrot fish actually eat away at the living coral and slender wrasses snap up the broken fragments. Fierce-looking moray eels hide in crevices and large sharks lurk in the deeper waters, hunting the big shoals of mackerel and tuna.

On the east of the New Territories of Hong Kong a number of deep sheltered bays protect the best coral reefs. Here, at Hoi Ha Wan, WWF are establishing a marine reserve. Over 50 species of coral can be found, including hard corals like the branching *Acropora pruinosa*, encrusting *Favia speciosa* and *Platygyra pini*, platelike *Pavona decussata* and *Lithophyllon*, and also some soft corals. Molluscs include beautiful small clams, cones, cowries and the colourful free-swimming nudibranchs. A multitude of bright coral fish sport among the reefs and the amazing poisonous-spined lionfish flash their fearsome display at potential enemies. Other invertebrates include beautiful sea lilies, spiny sea urchins and a wealth of colourful tube worms.

BELOW The sun starts to set behind the rugged peaks of Lantau Island, largest island in Hong Kong, seen across the beautiful expanse of the western harbour. OPPOSITE PAGE The rocky intertidal zone of Lamma Island is an ideal home to an array of molluscs, including stalked and acorn barnacles. All must be able to withstand rapid desiccation and heating under the tropical sun.

The warm waters of China's tropical coast shelter abundant marine life. ABOVE LEFT Dendrophylliid corals dangle their stinging tentacles to catch passing plankton on the gentle currents of a reef. ABOVE RIGHT Soft corals such as these grow to form beautiful marine 'gardens'. They lack the stony walls of the reef-building hard corals and often have a branching structure much like that of plants. BELOW LEFT A Spine-cheek Anemonefish (*Premnas biaculeatus*). Anemonefish, sometimes known as clownfish, are able to live among the tentacles of anemones without being stung because they are protected by a coating of mucus on their skins. BELOW RIGHT The Black-spotted Porcupine Fish (*Diodon hystrix*),

like other members of its family, is covered with spines. These lie flat when the fish is at rest, but if it puffs itself out in defence the body inflates like a ball and the spines stand erect. OPPOSITE PAGE, ABOVE Giant Clams (*Tridacna maxima*) embed themselves in the coral reef. They are harvested by fishermen for the meat in their strong muscles and for their shells, which are used to make lime. OPPOSITE PAGE, BELOW The Crown of Thorns Starfish (*Acanthaster plancii*) looks very decorative but it is a serious pest in some parts of the South China Sea where it infests and devours coral reefs.

Major Reserves in China

With over 600 reserves to choose from, it is no easy task to identify which ones are the most important or representative. Not all are open to the general public and some are very inaccessible.

In some places the huge size of the declared reserves gives them an apparent importance they may not merit on biological grounds whilst, at the other extreme, areas of supreme biological importance may not be protected at all, or have only one or two nominal protected areas that are quite insufficient in terms of regional conservation. The situation is also complicated by the fact that several different agencies are independently in the business of setting up nature reserves, and there is no central agency with complete information on the entire protected area system. Most of the reserves are under the management of the Ministry of Forestry, whilst a significant number are under the control of the national Environmental Protection Agency and others are under provincial or even municipal management.

However, in an attempt to bring together as much information as possible, in 1992 the World Wide Fund For Nature (WWF) organized a national review of protected areas in China, and held a workshop with experts from all provinces and relevant ministries to try to establish priorities. As a result, 40 reserves or reserve clusters, listed below, were identified as having global importance – though this may not represent the final answer to the problem. Many of these reserves were visited during the production of this book.

NORTH-EAST CHINA

The **Shuangtaizi** estuaries in Liaoning province are an 800-square-kilometre (300-square-mile) wetland area important for waterbirds and as a staging post for migrating cranes. They are also one of the only known breeding areas of the endemic Saunders's Gull.

Changbaishan covers an area of 1,908 square kilometres (737 square miles) in Jilin province, on the border with North Korea, and contains the highest mountain in north-east China. The reserve is one of the last localities of the Siberian Tiger in China, and is home to other animals such as Sable and Red Deer. It is important for protecting temperate forest systems and volcanic scenery.

The **Xianghai** and **Horqin** reserves are adjacent to each other in Jilin province and Inner Mongolia respectively. With a total area of 2,396 square kilometres (925 square miles), they encompass a system of lakes and marshes in the Kolsin steppelands. These wetlands are frequented by cranes and other important waterfowl, and rare vegetation is found on the sand dunes.

Huzhong and **Hanma** reserves straddle the Heilongjiang and Inner Mongolia boundaries, covering a total area of 3,292 square kilometres (1,271 square miles). These are the best examples of the northern taiga forests of China, with a variety of conifers and such northern animals as Moose, Wolverine and Hazel Grouse.

Zhalong reserve in Heilongjiang has an area of 2,100 square kilometres (810 square miles). These wetlands and reedbeds are the breeding grounds for several species of cranes, as well as a haven for many other waterbirds.

INNER MONGOLIA

Xilin Gele, 10,786 square kilometres (4,164 square miles), is a huge expanse of steppe grasslands. Mongolian Gazelle graze here and it is the largest grassland reserve in China.

Helanshan is a forested range of mountains along the Ningxia-Inner Mongolia border. This area of 2,255 square kilometres (870 square miles) is of geological interest as well as being important for Marco Polo Sheep and Blue Eared Pheasants.

NORTH CHINA

The **Huanghe** (Yellow river) delta is a major staging area for migrating cranes, ducks, geese, swans and many waders. In Shandong province 507 square kilometres (196 square miles) of the delta have been made into a reserve.

The **Qinling** mountains of Shaanxi province contain several nature reserves with a total area of 1,731 square kilometres (668 square miles). They provide protection for Giant Pandas, Golden Monkeys, Takin and other special Chinese fauna. The highest peak, Taibai, is a sacred mountain visited by pilgrims and others seeking peace in which to meditate.

Funiushan, in Henan province, comprises a cluster of small reserves centred on Laojunshan and Laojieling with a total area of 254 square kilometres (98 square miles). The region is important for preserving examples of transitional temperate and subtropical evergreen broadleaf forests.

The **Yancheng** marshes are a reserve of 467 square kilometres (180 square miles) on the east coast of Jiangsu province. The area is important for wetland birds and the Chinese Parrotbill. It is adjacent to the Dafeng reserve, where Père David's Deer have been reintroduced into the wild.

Shennongjia, with an area of 705 square kilometres (272 square miles) in Hubei province, is a rare example of extensive broadleaf and conifer forests within the Yangtze catchment. The reserve is famous for its Golden Monkeys, limestone pinnacles and the mythical 'wild man' of Hubei.

Shengjin lake in Anhui province is a reserve of 334 square kilometres (129 square miles). Like Poyang and Dongting, it is part of the lower Yangtze flood zone, and provides a wintering area for waterfowl. It is also important for freshwater fish species.

NORTH-WEST CHINA

Hanasi (or Kanas lake), 2,500 square kilometres (965 square miles), is the most important reserve in the Altai mountains of Xinjiang. The forests are of a northern coniferous type similar to those in the north-east, but there are many endemic varieties and spectacular glacial scenery.

THE TIBETAN PLATEAU AND HIMALAYAS

Qilianshan is a northerly temperate forest on the escarpment between Qinghai and Gansu provinces, and forms the north-east rampart of the Tibetan plateau. This reserve of 5,290 square kilometres (2,042 square miles) is primarily of botanical importance.

Qinghai lake in Qinghai province is the largest plateau lake in China and a notable breeding area for Bar-headed Geese, Cormorants and Brown-headed Gulls. Black-necked Cranes are autumn visitors. Only 533 square kilometres (206 square miles), a fraction of the total lake area, have been made a nature reserve.

Arjinshan is a huge reserve of 45,800 square kilometres (17,683 square miles) in the south-east corner of Xinjiang. It has a wide range of northern plateau habitats, including sandy deserts, cold deserts, mountains and salt lakes. Wildlife includes Yaks, Wild Ass, Tibetan Gazelles and Camels.

Tashikuorgan is another very large reserve in Xinjiang, covering 15,000 square kilometres (5,790 square miles). It is adjacent to the Pamir region of Afghanistan and Pakistan, and is important for Marco Polo Sheep, Snow Leopard and other high-altitude fauna and flora.

Jiangtang is a vast reserve of 240,000 square kilometres (92,664 square miles), lying in the north of Tibet in the centre of the Tibetan plateau. It preserves an enormous tract of cold deserts with numerous lakes and marshes and some glaciers. It is the home of Yaks, Snow Leopard and Tibetan Gazelles.

Zhufeng protects the highest regions in the world, being an area of 12,000 square kilometres (4,633 square miles) that rises to the peak of Everest and is contiguous with the Nepal Sagarmartha Park. Most of the reserve is rock and ice but some Himalayan wildlife occurs at the lowest altitudes.

Shenzha is a 30,000-square-kilometre (11,583-square-mile) site of marshy steppes and the breeding ground for Black-necked Cranes and other waterfowl. The reserve lies in the middle of Tibet on the plateau.

The **Xiaman** marshes in the north corner of Sichuan are China's largest peat swamp. As well as being the main breeding area of the Black-necked Crane, they are of importance for many wetland species. The reserve covers an area of 2,390 square kilometres (923 square miles).

There are several reserves packed into the **Nujiang-Lancangjiang** convergence on the Yunnan-Tibet border. Here, steep gorges separate the major rivers and an immense amount of biodiversity is crammed into a small area. The Grey Snub-nosed Monkey is endemic to this region. Together, the reserves total an area of 3,754 square kilometres (1,449 square miles).

SOUTH-WEST CHINA

The **Minshan** mountains of Sichuan are the main home of the Giant Panda population. About 3,000 square kilometres (1,158 square miles) of the area are protected in a series of rather spectacular reserves which contain other rare animals such as endemic pheasants, Takin and Golden Monkeys.

The **Qionglai** mountains are another Giant Panda stronghold in Sichuan on the opposite side of the Min river from Minshan. **Wolong** is the largest and most famous of four adjacent reserves with a total area of over 3,000 square kilometres (1,158 square miles). Red Pandas, Blue Sheep, and many endemic birds and plants add to the value of the region.

Haizishan covers 4,900 square kilometres (1,892 square miles) in west Sichuan. It protects some spectacular alpine scenery and high-altitude marshes which are the home of White-lipped Deer.

The **Ailao** and **Wuliang** mountain ranges in southern Yunnan form the home of the black Concolor Gibbon, Hume's Pheasant and a number of other important subtropical species. These neighbouring reserves have a combined area of 738 square kilometres (285 square miles).

CENTRAL CHINA

Fanjingshan is a famous reserve in Guizhou province, with an area of 419 square kilometres (162 square miles). This is the last home of the endemic Brelich's Snub-nosed Monkey and also protects Giant Salamanders and the beautiful Dove-tree.

Poyang is the largest freshwater lake in China. It lies in Jiangxi province and drains into the Yangtze river. A nature reserve of 224 square kilometres (86 square miles) on the west banks of the lake protects huge wintering populations of waterfowl, including the entire world population of Siberian Cranes.

Wuyishan, one of the highest mountains in south-east China, stands on the border between Jiangxi and Fujian provinces. This 601-square-kilometre (232-square-mile) Man and the Biosphere reserve is well known as an original collecting area for much of the region's endemic fauna. There is some spectacular scenery.

The **South China Tiger Area** is a whole range of mountains that straddle the provinces of Fujian, Jiangxi and Guangdong. Its total extent is over 1,000 square kilometres (386 square miles) and within it several small reserves protect core areas in a rugged and semi-wild landscape, where a few of the eponymous tigers still survive.

Dongting is one of the major Yangtze river lakes. It is in Hunan province and 1,843 square kilometres (711 square miles) of it have been made a reserve for wintering waterfowl, cranes, Scaly-sided Mergansers and Yangtze River Dolphins.

Nanling is a collection of small reserves that cross the Guangdong-Hunan border, covering some 500 square kilometres (193 square miles). **Babaoshan**, the most accessible of these, is the best-preserved subtropical evergreen forest in southern China and is full of important forest birds.

Nanshan Tiger Reserve is a wide conservation area spreading across the Guangdong, Jiangxi and Hunan borders. It is one of the last two areas where the South China Tiger still survives. Several reserves with a total of over 1,000 square kilometres (386 square miles) make up core areas of protection in a generally wild and rugged region.

Damingshan, with an area of 582 square kilometres (225 square miles), is the largest reserve in Guangxi. Part of the reserve is old logged forest but much of it, where there are steep limestone cliffs, is primary evergreen forest of great biological interest. Spectacular waterfalls and rare monkeys add to its value.

TROPICAL SOUTH CHINA

Xishuangbanna is Yunnan's premier tropical reserve, with a total of 2,418 square kilometres (934 square miles) in five separate sub-reserves. This is the best place in China to see Asian Elephants, Gaur and other tropical rainforest species.

Tongbiguan is another of Yunnan's tropical reserves. It is situated on the west side of the Nujiang (Salween) river and has some rather different species from Xishuangbanna. Here the Hoolock Gibbon calls, and this is the only reserve in China where sal forests, more typical of Myanmar and India, grow.

Nonggang forest is an area of karst limestone in Guangxi province, close to the border with Vietnam. It is famous as the home of two endemic leaf monkeys and has the potential to be developed as a transfrontier reserve with adjacent areas in Vietnam.

Dongzhaigang and **Qinlangang** are both mangrove reserves on the east side of Hainan Island. They have a combined area of only 54 square kilometres (21 square miles) but are the largest and best-preserved examples of mangroves in China. They are important staging and wintering areas for waders.

South-west Hainan is an area of extreme biological interest, surprisingly rich in evergreen plants and with several endemic animal species such as White-eared Partridge. The main protected areas here are Jianfengling, Bawangling and Wuzhishan, totalling 1,383 square kilometres (534 square miles).

Bibliography

Geography

Bonavia, J. (1985) *An Illustrated Guide to the Yangzi River*. The Guidebook Company, Hong Kong.

Chan, V. (1994) *Tibet Handbook*. Moon Publications Inc, Chico, California.

Goldstein, M.C. and Beall, C.M. (1990) *Nomads of Western Tibet*. Odyssey Productions Ltd, Hong Kong.

Holdsworth, M. (1993) *Sichuan*. Odyssey, Hong Kong.

Holledge, S., Hunt, J., Rolnick, H. and Courtauld, C. (1989) *All China – An Illustrated Introduction to China, its People and Culture*. Passport Books, Hong Kong.

Sun Jingzhi (1988) *The Economic Geography of China*. Oxford University Press, Hong Kong.

Zhao Songqiao (1986) *Physical Geography of China*. Science Press, Beijing and John Wiley & Sons Inc, New York.

General Natural History

Dudgeon, D. and Corlett, R. (1994) *Hills and Streams – An Ecology of Hong Kong*. Hong Kong University Press.

Guangdong Forestry Department and S.E. China Endangered Species Institute (1987) *Guangdong Wildlife and Colour Illustrations* (in Chinese). Guangdong Technology Press, Guangzhou.

Hill, D.S. and Phillipps, K. (1981) *A Colour Guide to Hong Kong Animals*. The Government Printer, Hong Kong.

Laidler, L. and K. (1996) *China's Threatened Wildlife*. Blandford Press, London.

Lee Fen-Lan and Yang Hui-Lang (1993) *Nature Conservation in Taiwan, ROC*. Council of Agriculture, Executive Yuan, Taipei.

Li Wenhua and Zhao Xianying (1989) *China's Nature Reserves*. Foreign Languages Press, Beijing.

Tang Hsiao-yu (ed) (1989) *Nature Protection*. Taiwan Government Agricultural Department Press, Taipei.

Tang Xiyang (1987) *Living Treasures – An Odyssey through China's Extraordinary Nature Reserves*. Bantam Books, New York.

Thrower, S.L. (1988) *Hong Kong Trees*. The Urban Council, Hong Kong.

Wilson, E.H. (1913) *A Naturalist in Western China*. Methuen & Co Ltd, London.

Zhao Ji, Zheng Guangmei, Wang Huadong and Xu Jialin (1990) *The Natural History of China*. Collins, London.

Conservation and Reserves

Ashworth, J.M., Corlett, R.T., Dudgeon, D., Melville, D.S. and Tang, W.S.M. (1993) *Hong Kong Flora and Fauna: Computing Conservation*. World Wide Fund For Nature Hong Kong.

Biodiversity Committee of the Chinese Academy of Sciences (1992) *Biodiversity in China – Status and Conservation Needs*. Science Press, Beijing.

Changchun Institute of Geography (1990) *The Conservation Atlas of China*, Science Press, Beijing.

Fang Donghan and Zhen Yaozu (1990) *The Taibai Mountain*. Shaanxi Peoples' Press.

Guizhou Provincial Committee for Conservation (1982) *Scientific Survey of Fangjingshan Mountain Reserve*. Guizhou Province Bureau of Conservation, Guiyang.

National Environmental Protection Agency (1992) *Nature Reserves in China*. China Environmental Science Press, Beijing.

National Environmental Protection Agency (1994) *Environmental Action Plan of China 1991 – 2000*. China Environmental Science Press, Beijing.

National Environmental Protection Agency (1994) *China – Biodiversity Conservation Action Plan*. Beijing.

State Council of People's Republic of China (1994) *China's Agenda 21*. China Environmental Science Press, Beijing.

Wang Jingcao (1986) *Science Investigation in Natural Conservation of Guniujiang* (in Chinese). China Zhanwan Press.

Wang Sung and MacKinnon, J. (1993) Urgent recommendations to save China's biological diversity. *Chinese Biodiversity*, Vol I:2-13.

Yunnan Forestry Bureau (1987) *Yunnan Nature Reserves* (in Chinese). China Forestry Press, Beijing.

Zhao Kelin (1985) *Scientific Investigations of Xishuangbanna Nature Reserve* (in Chinese). Yunnan Technology Press, Kunming.

Mammals

Corbet, G.B. and Hill, J.E. (1980) *A World List of Mammalian Species*. British Museum, London.

Corbet, G.B. and Hill, J.E. (1992) *The Mammals of the Indomalayan Region: A Systematic Review*. Oxford University Press, Oxford.

Feng Zuo-jian, Cai Gui-guan and Zheng Chang-lin (1986) *The Mammals of Xizang* (in Chinese). Science Press, Beijing.

Hu Jinchu and Wang Xizhi (1984) *Sichuan Fauna Economica Volume 2: Mammals* (in Chinese). Sichuan Science and Technology Press, Chengdu.

Ma Yong, Wang Fenggui, Jin Sangke and Li Sihua (1987) *Glires (Rodents and Lagomorphs) of Northern Xinjiang and their Zoogeographical Distribution* (in Chinese). Science Press, Academia Sinica, Beijing.

Wang Sung, Wang Jia-jun and Luo Yi-nin (1994) *Worldist of Mammalian Names* (Latin, Chinese and English). Science Press, Beijing.

Giant Panda

Campbell, J.J.N. and Qin Zisheng (1983) Interaction of Giant Pandas, bamboo and people. *Journal of the American Bamboo Society*, Vol 4:1-34.

Hu Jinchu (1981) *Ecology and Biology of the Giant Panda, Golden Monkey and Takin* (in Chinese). Sichuan People's Publishing House, Chengdu.

Johnson, K., Schaller, G.B. and Hu Jinchu (1988) Responses of Giant Pandas to a bamboo die-off. *National Geographic Journal*.

Joint Survey Team (Ministry of Forestry/World Wide Fund For Nature) (1989) *Final Report of the Joint Giant Panda Survey Team* (in Chinese). 2 Vols. Mimeo.

Ministry of Forestry – World Wide Fund For Nature (1989) *National Conservation Management Plan for the Giant Panda and its Habitat Draft*. China Alliance Press, Hong Kong.

Pan Wenshi, Lu Zhi (1988) *The Giant Pandas in the Qinling Mountains*. Beijing University Press.

Schaller, G.B., Hu Jinchu, Pan Wenshi and Zhu Jing (1985) *The Giant Pandas of Wolong*. The University of Chicago Press, Illinois.

Schaller, G.B., Teng, Q., Johnson, K., Wang, X., Shen, H. and Hu Jinchu (1989) Feeding ecology of Giant Panda and Asiatic Black Bear in Tangjiahe Reserve, China. In: Gittleman, J. (ed) *Carnivore Behaviour, Ecology and Evolution*. Cornell University Press, Ithaca, New York.

Birds

Chen Tso-hsin (1987) *A Synopsis of the Avifauna of China*. Science Press, Beijing.

De Schaunesee, R.M. (1984) *The Birds of China*. Oxford University Press, Oxford.

Etchécopar, R.D. and Hüe, F. (1978, 1982) *Les Oiseaux de Chine, de Mongolie et de Corée*. 2 Vols. Editions du Pacifique, Papeete, Tahiti.

Fan Zhongming (ed) (1990) *Chinese Birds Species Summary* (in Chinese). Liaoning Science and Technology Press.

King, B., Woodcock, M. and Dickinson, E.C. (1975) *A Fieldguide to the Birds of South-east Asia*. Collins, London.

Monroe, B.L. Jr, and Sibley, C.G. (1993) *A World Checklist of Birds*. Yale University Press, New Haven and London.

Numata, M. (ed) (1974) *The Flora and Vegetation of Japan*. Elsevier Scientific, New York.

Qian Yanwen and China Wildlife Protection Association (eds) (1995) *Atlas of Birds of China*. Henan Science and Technology Press.

Viney, C., Phillipps, K. and Lam Chiu-ying (1994) *Birds of Hong Kong and South China*. Government Printer, Hong Kong.

Wang Zijiang (ed) (1991) *A Manual for the Identification of Precious and Rare and Common Birds in Yunnan Province, China* (in Chinese). Yunnan University Press.

Invertebrates

Li Chuanglong and Zhu Baoyun (1992) *Atlas of Chinese Butterflies*. Shanghai Fareastern Press.

Plants and Vegetation

Fu Li-kuo (ed) (1992) *China Plant Red Data Book – Rare and Endangered Plants Volume 1*. Science Press, Beijing.

Mierow, D. and Shrestha, T.B. (1978) *Himalayan Flowers and Trees*. Sahayogi Prakashan, Kathmandu.

Taylor, A.H. and Qin Zisheng (1988) Regeneration patterns in old growth Abies-Tsuga forest in the Wolong Natural Reserve, Sichuan. *China Journal of Ecology*.

Wang Chi-wu (1961) *The Forests of China with a Survey of Grassland and Desert Vegetation*. Harvard University Press, Cambridge, Massachusetts.

Wolong Nature Reserve Management Administration (1987) *Wolong Vegetation and Plant Resources* (in Chinese). Sichuan Science and Technology Press, Chengdu.

Xu Zaifu, Tao Guoda and Tang Jiakun (1989) *Tropical Wild Flowers and Plants in Xishuangbanna*. Agricultural Press, Kunming.

Yunnan Institute of Tropical Botany (1983) *List of Plants in Xishuangbanna*. Yunnan Minority Press, Kunming.

Zhang Jingwei (ed) (1982) *The Alpine Plants of China*. Science Press, Beijing.

Zhang Qitai, Feng Zhizhou and Yang Zenghong (1988) *Rare Flowers and Unusual Trees*. China Esperanto Press, Beijing.

Marine

Scott, P.J.B. (1984) *The Corals of Hong Kong*. Hong Kong University Press.

State Oceanic Administration (1993) *Action Plan for Marine Biodiversity Protection in China* (in Chinese). Mimeo.

Tai Chang-fung (1991) *Study of Coral Resources Around Nuclear Power Stations* (in Chinese). Oceanic Institute of Taiwan University, Taipei.

Wang, C.X. (1981) Studies of the fish fauna of the South China Sea. *Hong Kong Fisheries Bulletin*. 2:41-50.

Wetlands

Irving, R. and Morton, B. (1988) *A Geography of the Mai Po Marshes*. Hong Kong University Press.

Lu Jianjian (1990) *Wetlands in China* (in Chinese). East China Normal University Press, Shanghai.

Scott, D.A. (1989) *A Directory of Asian Wetlands*. IUCN, Cambridge, UK.

Wang Baojin, Wang Haibing, Song Xiangjin, Melville, D.S. and MacKinnon, J. (1993) *Jiangxi Poyang Lake National Nature Reserve Management Plan*. China Forestry Publishing House, Beijing.

Index